AF748074

Basic and Clinical Understanding of Microcirculation

Edited by Kaneez Fatima Shad, Seyed Soheil Saeedi Saravi and Nazar Luqman Bilgrami

Published in London, United Kingdom

IntechOpen

Supporting open minds since 2005

Basic and Clinical Understanding of Microcirculation
http://dx.doi.org/10.5772/intechopen.77660
Edited by Kaneez Fatima Shad, Seyed Soheil Saeedi Saravi and Nazar Luqman Bilgrami

Contributors
Victoria Fateeva, Vorob'va Olga Vladimirovna, Nikulina Ksenia Vladimirovna, Kristina Konstantinovna Khacheva, Khakimova Gulnara Rinatovna, Epstein Oleg Ilyich, Manuel F Navedo, Madeline Nieves, Arsalan U. Syed, Thanhmai T Le, Andrey Fedorovich, Saadat Huseynova, Samaya Alizada, Jamila Kurbanova, Afat Hasanova, Nushaba Panakhova, Jefferson Frisbee, Brayden Halvorson, Gausal Khan, Indranil Biswas, Wissam AbouAlaiwi, Sidney Ley, Kazuo Maruyama, Junko Maruyama, Hirofumi Sawada

© The Editor(s) and the Author(s) 2020
The rights of the editor(s) and the author(s) have been asserted in accordance with the Copyright, Designs and Patents Act 1988. All rights to the book as a whole are reserved by INTECHOPEN LIMITED. The book as a whole (compilation) cannot be reproduced, distributed or used for commercial or non-commercial purposes without INTECHOPEN LIMITED's written permission. Enquiries concerning the use of the book should be directed to INTECHOPEN LIMITED rights and permissions department (permissions@intechopen.com).
Violations are liable to prosecution under the governing Copyright Law.

CC BY

Individual chapters of this publication are distributed under the terms of the Creative Commons Attribution 3.0 Unported License which permits commercial use, distribution and reproduction of the individual chapters, provided the original author(s) and source publication are appropriately acknowledged. If so indicated, certain images may not be included under the Creative Commons license. In such cases users will need to obtain permission from the license holder to reproduce the material. More details and guidelines concerning content reuse and adaptation can be found at http://www.intechopen.com/copyright-policy.html.

Notice
Statements and opinions expressed in the chapters are these of the individual contributors and not necessarily those of the editors or publisher. No responsibility is accepted for the accuracy of information contained in the published chapters. The publisher assumes no responsibility for any damage or injury to persons or property arising out of the use of any materials, instructions, methods or ideas contained in the book.

First published in London, United Kingdom, 2020 by IntechOpen
IntechOpen is the global imprint of INTECHOPEN LIMITED, registered in England and Wales, registration number: 11086078, 7th floor, 10 Lower Thames Street, London, EC3R 6AF, United Kingdom
Printed in Croatia

British Library Cataloguing-in-Publication Data
A catalogue record for this book is available from the British Library

Additional hard and PDF copies can be obtained from orders@intechopen.com

Basic and Clinical Understanding of Microcirculation
Edited by Kaneez Fatima Shad, Seyed Soheil Saeedi Saravi and Nazar Luqman Bilgrami
p. cm.
Print ISBN 978-1-78985-551-7
Online ISBN 978-1-78985-552-4
eBook (PDF) ISBN 978-1-83962-968-6

For EU product safety concerns:
IN TECH d.o.o., Prolaz Marije Krucifikse Kozulić 3, 51000 Rijeka, Croatia, info@intechopen.com or visit our website at intechopen.com.

We are IntechOpen, the world's leading publisher of Open Access books

Built by scientists, for scientists

4,900+
Open access books available

124,000+
International authors and editors

140M+
Downloads

151
Countries delivered to

Our authors are among the
Top 1%
most cited scientists

12.2%
Contributors from top 500 universities

CLARIVATE ANALYTICS
BOOK CITATION INDEX
INDEXED

WEB OF SCIENCE™

Selection of our books indexed in the Book Citation Index
in Web of Science™ Core Collection (BKCI)

Interested in publishing with us?
Contact book.department@intechopen.com

Numbers displayed above are based on latest data collected.
For more information visit www.intechopen.com

Meet the editors

Professor Kaneez Fatima Shad, a Pakistani-born Australian neuroscientist with a medical background, received a PhD in 1994 from the Faculty of Medicine, UNSW, Australia, followed by a post doc at the Allegheny University of Health Sciences, Philadelphia, USA. She taught medical and biological sciences at various universities in Australia, the United States, United Arab Emirates, Bahrain, Pakistan, and Brunei. During this period, she was also engaged in research supported by local and international grants (total of over 2.3 million USD) and translating it into products such as a rapid diagnostic test for stroke and other vascular disorders such as schizophrenia. She published over 60 articles in refereed journals, edited 4 books, wrote 7 book chapters, presented at over 75 international conferences, and mentored 34 postgraduate students.

Dr. Seyed Soheil Saeedi Saravi, PharmD, PhD, senior scientist at Harvard Medical School, received his PhD in 2016 from Tehran University of Medical Sciences, followed by a post doc at Harvard Medical School. His research areas include vascular biology and atherosclerosis. He has authored over 70 peer-reviewed publications, edited 9 books and chapters, and presented at over 25 international conferences. He has served on the editorial board of 20 prestigious journals, and is a professional member of over 10 international scientific associations, including American Heart Association and European Society of Cardiology. He has been honored with a Harvard fellowship, SwissLife stiftung, first-ranked award at national PhD entrance exam, and national excellent medical student, young thesis editor, young researcher, and National Elites Foundation awards.

Dr. Nazar Luqman Bilgrami is an experienced cardiologist with over 30 years of clinical, research and academic experience. He was trained in India, United Kingdom and Australia. Having worked previously in Brunei Darussalam as a cardiologist and adjunct associate professor, he has now moved to Australia. At present he is working with Austin Health in Melbourne, Australia. He is an accomplished researcher with several publications. In addition to his clinical work, he is currently pursuing research in cardiovascular issues related to heart failure. His contribution as a co-editor has gone a long way in compiling this book on microcirculation and making it a worthy reference book for cardiologists.

Contents

Preface

Microcirculation occurs through an extensive network of vasculature spread throughout the body. An intact microcirculatory system is a prerequisite for adequate perfusion and hence normal functioning of organs in the body. Microcirculatory disorders could result from decreased microcirculation, damage to the endothelial cells, altered vascular permeability and/or changes in microvascular reactivity. Early identification of microcirculatory impairment through investigative means could help improve outcome for a large number of disease states.

This book entitled "Basic and Clinical Understanding of Microcirculation" effectively develops the very basic concepts of microvascular circulation. The first chapter describes the role of ion channels in regulating vascular smooth muscles. Subsequent chapters cover topics such as how diseases like diabetes affect the regulation of cerebrovascular tone and lead to neurological disorders. Later, the role of endothelial dysfunction, relation to nitric oxide in maintaining vasodilation, and abnormalities in disease states are discussed. Reading these chapters on the basic aspects also makes one understand how vascular dysfunction is related to the diseases like diabetes and atherosclerosis.

Clinical section of the book begins with describing the pivotal role of endothelial dysfunction in cardiovascular diseases, followed by association with the peripheral blood inflammatory markers and their role in depression. Similarly, when the active monocellular layer of cells gets affected, it could lead to atherosclerosis, thrombosis, diabetes, hypertension etc.

In this book the authors thoroughly describe methods to evaluate microcirculation using modern diagnostic technology like laser doppler flowmetry. Last but not least, the readers will be intrigued with how microvascular circulation could be related to growth and development of newborn. I hope hhat those who read this book will benefit through better understanding of the various facets of microcirculation.

I would like to thank my co-editors Nazar Luqman Bilgrami (Cardiologist at Austin Health, Melbourne, Australia) and Seyed Soheil Saeedi Saravi (Cardiovascular Division, Harvard Medical School, Boston, USA) for their contribution towards compiling this book, editing the chapters, and writing the preface.

Production of this book would not have been possible without the contribution from the experts in the field and the continuous hard work of the Author Service Manager and Commissioning Editor at IntechOpen.

Finally, I would like to emphasise that this book is meant for a broad range of readers including undergraduates, graduates, researchers, teachers, medical professionals, and cardiovascular specialists.

Kaneez Fatima Shad
Professor,
School of Life Sciences,
Faculty of Science,
Centre for Health Technologies,
University of Technology Sydney,
Australia

Seyed Soheil Saeedi Saravi
Harvard Medical School,
USA

Nazar Luqman Bilgrami
Austin Health,
Melbourne, Australia

Section 1

Basics of Microcirculation

Chapter 1

Ion Channels and Their Regulation in Vascular Smooth Muscle

Arsalan U. Syed, Thanhmai Le, Manuel F. Navedo and Madeline Nieves-Cintrón

Abstract

Vascular smooth muscle excitability is exquisitely regulated by different ion channels that control membrane potential (E_m) and the magnitude of intracellular calcium inside the cell to induce muscle relaxation or contraction, which significantly influences the microcirculation. Among them, various members of the K^+ channel family, voltage-gated Ca^{2+} channels, and transient receptor potential (TRP) channels are fundamental for control of vascular smooth muscle excitability. These ion channels exist in complex with numerous signaling molecules and binding partners that modulate their function and, in doing so, impact vascular smooth muscle excitability. In this book chapter, we will review our current understanding of some of these ion channels and binding partners in vascular smooth muscle and discuss how their regulation is critical for proper control of (micro)vascular function.

Keywords: ion channels, signal transduction, vascular tone, vascular smooth muscle, microcirculation

1. Introduction

Vascular smooth muscle cells wrapping around small resistance arteries and arterioles are crucial for vascular reactivity [1]. These cells enable dynamic, moment-to-moment control of vessel diameter and pressure-induced contraction (e.g., vascular tone). This control is central to autoregulation of resistance vessels, maintenance of vessel caliber independently of changes in blood pressure, and proper perfusion to meet the metabolic demands of a given tissue.

To regulate arterial diameter, vascular smooth muscle receives and integrates many inputs, including changes in intraluminal pressure, vasoconstrictor and vasodilatory signals from endothelial cells lining the inner arterial wall, and nerve terminals innervating the vessels [2]. These inputs regulate vascular smooth muscle excitability, at least in part, by modulating the activity of a number of ion channels to control membrane potential (E_m) and the magnitude of intracellular Ca^{2+} concentration ($[Ca^{2+}]_i$) [1]. Among the many ion channels, transient receptor potential (TRP) channels, voltage-gated (K_V), Ca^{2+}-activated (BK_{Ca}) and inward rectifier (K_{ir}) K^+ channels, and voltage-gated Ca^{2+} channels (VGCC) are fundamental in transducing mechanical force, establishing E_m, and regulating $[Ca^{2+}]_i$ [1]. Mechanisms for the regulation of vascular smooth muscle ion channels, including those mentioned above, involve agonist-independent and agonist-dependent activation of G_q and/or G_s protein-coupled receptors (G_xPCR) [3]. The optimal activation and regulation of these mechanisms are often dependent on the formation of macromolecular complexes driven by scaffold proteins to target and compartmentalize proteins/signaling events to specific regions and substrates [4].

In this chapter, we provide a brief overview of our current understanding of the role of different subunits of TRP channels, several K^+ channel subtypes, as well as key VGCCs in control of vascular smooth muscle excitability and (micro)vascular function. We will also discuss their regulation by signaling pathways and macromolecular complexes. Note that many more studies than those cited here can be found in the scientific literature. Comprehensive studies exploring additional aspects of vascular ion channel regulation, vascular smooth muscle excitability, and (micro) vascular function in health and disease can be found in recent reviews [1, 5–9].

2. TRP channels

The TRP channel superfamily is composed of 28 members divided into 6 subfamilies based on their molecular and biophysical properties [10]. Functional TRP channels are composed of four subunits, each with six membrane-spanning helices that can exist in a homomeric or heteromeric form. Vascular smooth muscle cells express a number of these channels, including members of the TRPC, TRPV, TRPM, and TRPP subfamilies [7, 9]. The function of TRP channels in vascular smooth muscle ranges from regulating contractility to modulating the proliferative state of the cells. In this section, we focused our discussion on the functional role and regulation of specific TRP channels in vascular smooth muscle excitability.

2.1 Functional role of vascular smooth muscle TRP channels

The function of different TRP channels in vascular smooth muscle has been unmasked using conventional and innovative molecular, pharmacological, and genetic approaches. These approaches revealed distinctive roles of specific TRP channel subfamilies in modulating vascular smooth muscle excitability and vascular reactivity. For instance, studies have found that TRPC3 channels are not essential for pressure-induced constriction. Yet, TRPC3 channels play a key role in receptor-mediated vasoconstriction of resistance arteries upon activation of various G_qPCRs, including purinergic receptors, endothelin (ET_A) receptors, and angiotensin II (AT_1) receptors [7, 11, 12]. The mechanisms for TRPC3 channel activation involve direct coupling of the channel with IP_3 receptors located in the sarcoplasmic reticulum (SR) in a process that does not require SR Ca^{2+} release [13]. TRPC5 channels, in association with TRPC1 channels, have been shown to contribute to store-operated Ca^{2+} entry in vascular smooth muscle from arterioles [14], which may modulate cell excitability. Expression of TRPP channels in vascular smooth muscle contributes to stretch-activated cation currents, which causes vascular smooth muscle membrane depolarization. The activity of these channels has been associated with stretch-dependent regulation of vascular tone in cerebral arteries and control of systemic blood pressure [15, 16]. The nonselective cation channel TRPC6 is involved in Ca^{2+} mobilization leading to vascular smooth muscle contraction [17]. More recently, these channels have been shown to be part of a mechanosensation complex [18]. In this model (see **Figure 1**), stretch is "sensed" by AT_1 receptors. These receptors then induce the activation of TRPC6 channels to bolster Ca^{2+} release from the SR, which triggers TRPM4 channel activity and vascular smooth muscle contraction. This model reveals an exquisite and finely orchestrated macromolecular complex for control of stretch-induced contraction.

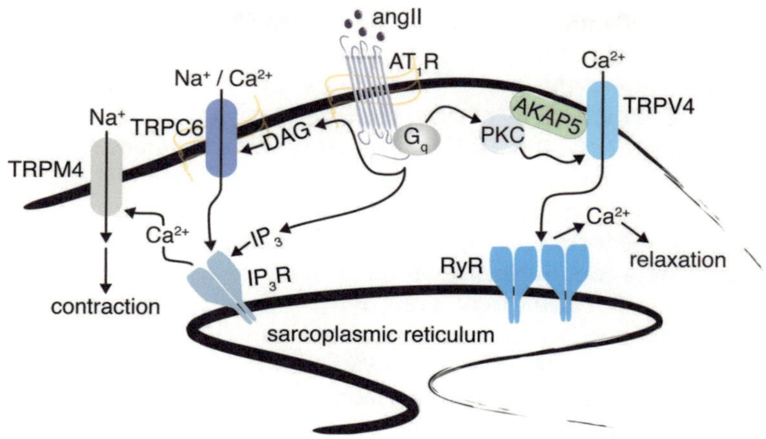

Figure 1.
Regulation of vascular smooth muscle excitability by TRP channels. Diverse TRP channels contribute to vascular smooth muscle relaxation or contraction. Ca^{2+} influx via TRPV4 channels is associated with an increase in the frequency of Ca^{2+} release through ryanodine receptors (RyR) located in the sarcoplasmic reticulum (e.g., Ca^{2+} sparks). The production of Ca^{2+} sparks triggers the activity of adjacent BK_{Ca} channels at the surface membrane to promote K^+ efflux, membrane potential hyperpolarization, and vascular smooth muscle relaxation [18]. TRPV4 channel activity can be stimulated by activation of AT_1 receptors (AT_1R) by angiotensin II (angII) via an AKAP5-mediated PKC signaling pathway [20, 25]. This AT_1R/AKAP5/PKC/TRPV4 axis may serve as a negative feedback mechanism to offset stimulation of L-type Ca^{2+} channel $Ca_V1.2$ (LTCCs) leading to contraction in response to angII. AT_1R, independent of angII, has also been implicated in a mechanosensitive pathway that activates TRPC6 channels to boost cytosolic Ca^{2+} via stimulation of IP_3 receptors (IP_3R). This increase in cytosolic Ca^{2+} promotes the activity of TRPM4 channels leading to vascular smooth muscle membrane depolarization and subsequent contraction in response to stretch [17]. The orange wavelike lines highlight stretch-sensing proteins. The model was generated by taking in consideration studies cited and described above. Na^+ = sodium; Ca^{2+} = calcium; AKAP5 = A kinase anchoring protein 5; PKC = protein kinase C; DAG = diacylglycerol; IP_3 = inositol trisphosphate.

Contrary to the contractile influences of the TRP channels described above, activation of TRPV4 channels is associated with vascular smooth muscle relaxation (**Figure 1**) [19]. The underlying mechanism involves the formation of a macromolecular complex between TRPV4 and BK_{Ca} channels in the plasma membrane and ryanodine receptors (RyR) in the SR. This complex does not depend on direct protein-protein interactions between TRPV4 and BK_{Ca} channels and RyR but requires their close physical proximity. The close association between these proteins facilitates Ca^{2+} influx via TRPV4 channels that stimulates RyR activity resulting in the generation of Ca^{2+} sparks. The ensuing Ca^{2+} sparks activate BK_{Ca} channels leading to vascular smooth muscle membrane potential hyperpolarization and relaxation [20]. Because of the high Ca^{2+} permeability of TRPV4 channels in vascular smooth muscle [21], they could potentially activate adjacent BK_{Ca} channels directly to regulate cell excitability and vascular reactivity. Examination of this exciting possibility is an excellent opportunity for further research. Moreover, additional work is needed to comprehensively define the role of TRP channels in vascular smooth muscle in the microcirculation during physiological and pathological conditions.

2.2 Regulation of TRP channels in vascular smooth muscle

It has been well documented that agonist-dependent activation of G_qPCRs/protein kinase C (PKC) and G_sPCRs/protein kinase A (PKA) signaling regulates TRP channels [10]. Protein kinase G (PKG) modulation of TRP channels has also been reported. However, how these signaling pathways control TRP channel activity in vascular smooth muscle, as well as the underlying consequences in vascular reactivity are not well understood. A limited number of studies have shown that PKG inhibits TRPC3 channels and that this may contribute to nitric oxide (NO)-mediated relaxation, but the mechanisms require further examination [22]. PKA signaling stimulates the activity of several members of the TRPV channel subfamily while inhibiting TRPC5 and TRPC6 channels in vascular smooth muscle [7]. PKA-mediated inhibition of TRPC6 channels in response to agonist stimulation was shown to be dependent on phosphorylation of the TRPC6 subunit at threonine 69, which resulted in a reduction in angiotensin II (angII)-induced vasoconstriction [23], thus revealing a comprehensive mechanism for the regulation of TRP channels and the underlying effects in vascular reactivity.

PKC activity inhibits TRPC3 channels and activates TRPM4 and TRPV4 channels in smooth muscle [7, 9, 21, 24]. TRPM4 activation by PKC proceeds, at least in part, by stimulating channel trafficking and membrane translocation via a mechanism requiring PKCδ [25]. This PKCδ-dependent anterograde TRPM4 trafficking is functionally relevant as it promotes vascular smooth muscle contraction. PKC-dependent regulation of TRPV4 channels has been suggested to be critical for counteracting the vasoconstriction stimulated by angII [21]. Intriguingly, TRPV4 channels are found in complex with the scaffold A kinase anchoring protein 5 (AKAP5 = human AKAP79 and murine AKAP150) in vascular smooth muscle (**Figure 1**) [21, 26]. This scaffold protein provides a platform for targeting and compartmentalization of signaling molecules (e.g., PKA, PKC, protein phosphatase 2B = PP2B) to specific substrates (e.g., ion channels) [4]. With a suggested distance between them of ~200 nm [26], optimal AKAP5-anchored PKC modulation of TRPV4 activity is highly dependent on the distance between the targeted kinase and the ion channel. This tight regulation of TRPV4 activity by AKAP5 may be essential for vascular smooth muscle excitability given the high Ca^{2+} permeability of these channels, as mentioned previously. Mechanisms regulating TRPP channels by PKA and/or PKC in vascular smooth muscle are currently unclear. The studies discussed above reveal the complex functional role of TRP channels in vascular smooth muscle and how their regulation alters vascular function and highlights unique opportunities for further research. For instance, it will be important to define the role of the AKAP5/PKC/PKA complex and its association with other TRP channels in regulating vascular smooth muscle excitability and vascular reactivity. It also remains to be determined whether dynamic trafficking of other TRP channels and the involvement of the AKAP5/PKC/PKA complex in this process also play a role in fine-tuning vascular reactivity.

3. K^+ channels

The activity of K^+ channels determines vascular smooth muscle membrane potential and is therefore key regulators of vascular tone [1]. By setting and controlling the membrane potential, these channels influence the levels of $[Ca^{2+}]_i$ and therefore, vascular smooth muscle contraction. Intriguingly, a subset of K^+ channels has also been linked to regulation of vascular smooth muscle proliferation (see recent review on this topic in [27, 28]). Vascular smooth muscle cells express a wide variety of isoforms from several classes of K^+ channels, including K_V, BK_{Ca}, and K_{ir} channels (see **Figure 2**). In the following sections, we will describe the expression, function, and regulation of these K^+ channels and their control of vascular function.

3.1 K_V channels

A number of K_V channels isoforms are expressed in vascular smooth muscle, including members of the K_V1 ($K_V1.1$, $K_V1.2$, $K_V1.3$, $K_V1.5$, $K_V1.6$), K_V2 ($K_V2.1$), and

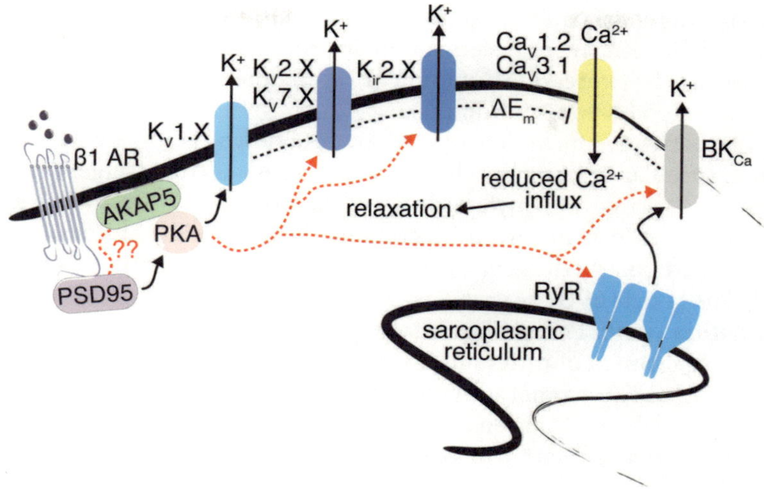

Figure 2.
Regulation of vascular smooth muscle excitability by K+ channels. Vascular smooth muscle cells express a number of K^+ channels that, upon activation, provide negative feedback regulation of membrane potential and vascular tone. K^+ channel activation results in efflux of K^+ ions that hyperpolarize the membrane potential (ΔE_m; denoted by dotted black lines) leading to a reduction in the open probability of LTCC $Ca_V1.2$ and TTCC $Ca_V3.1$ that results in a decrease in global $[Ca^{2+}]_i$ leading to vascular smooth muscle relaxation [32, 95]. As mentioned in ***Figure 1****, activation of ryanodine receptors (RyR) K^+ channels can be distinctively regulated by vasoactive agents acting through the G_s/AC/PKA, NO/sGC/PKG, and G_q/PLC/PKC axes to either increase or decrease their activity. An example is the regulation of $K_V1.X$ channels by β adrenergic signaling (e.g., through β1 adrenergic receptors) [44–46]. Here, $K_V1.X$ channel regulation by β1 adrenergic signaling requires PKA targeting to the channel by postsynaptic density protein 95 (PSD95). Whether PSD95 as well as further scaffolding of proteins by AKAP5 is required for PKA regulation of $K_V1.X$ and other K^+ channels (e.g., $K_V2.X$, $K_V7.X$, $K_{ir}2.X$) as well as RyR is unclear (denoted by dotted red lines in the cartoon). The model was generated by taking in consideration studies cited and described above. K^+ = potassium; Ca^{2+} = calcium; AKAP5 = A kinase anchoring protein 5; BK_{Ca} = large-conductance Ca^{2+} activated potassium channels; AC = adenylyl cyclase; PKA = protein kinase A; NO = nitric oxide; sGC = soluble guanylyl cyclase; PKG = protein kinase G; PLC = phospholipase C; PKC = protein kinase C.*

K_V 7 (K_V7.1–5), as well as several silent K_V subunits (e.g., K_V9.3) [1]. K_V channels are formed by a homo- or heterotetrameric assembly of pore-forming α subunits and regulatory β subunits. Key examples with functional relevance in vascular smooth muscle excitability from multiple vascular beds are K_V channels assembled by K_V1.2/K_V1.5 subunits, K_V2.1 subunits with K_V9.3, and K_V1/K_V7 subunits. These subunit compositions confer functional and pharmacological diversity essential for fine-tuning vascular smooth muscle excitability and vascular reactivity [29–31]. The mechanisms for this involve activation of K_V channels by a depolarizing stimulus (e.g., stretch-induced depolarization) and their regulation by vasoactive agents [1, 32, 33]. The resultant K^+ efflux hyperpolarizes the membrane potential of vascular smooth muscle leading to a decrease in the open probability of voltage-gated L-type Ca^{2+} channels Ca_V1.2 (LTCCs), which contributes to decrease $[Ca^{2+}]_i$ and relaxation (**Figure 2**). Conversely, their inhibition depolarizes the membrane potential, which will increase the open probability of LTCCs and lead to an increase in global $[Ca^{2+}]_i$ and vascular smooth muscle contraction. This negative feedback regulation, together with other K^+ channels, is essential for fine control of vascular smooth muscle excitability and vascular reactivity. However, in the context of the microcirculation, not much is known regarding the expression, functional composition, physiological role, and regulation of K_V channels in arteriolar vascular smooth muscle. Consideration of these issues is important as recent studies have implicated that impairment in K_V channel expression and/or function in the development of channelopathies are associated with small vessel diseases [34]. Moreover, additional pathologies such as hypertension, metabolic disorders, and diabetic hyperglycemia impair (micro)vascular function, at least in part, by altering the expression/function of K_V channels, but mechanisms remain not fully understood (see review in [5]).

Many vasoactive agents modulate vascular function by acting on K_V channels expressed in vascular smooth muscle. For instance, agents that trigger activation of G_qPCRs, such as angII, phenylephrine, and endothelin 1, are known to stimulate vasoconstriction, at least in part, by decreasing the expression and/or function of K_V channels, particularly those of the K_V1, K_V2, and K_V7 subfamilies (**Figure 2**) [1, 35–37]. The effects of these vasoconstrictors are related to PKC-mediated phosphorylation and/or changes in surface expression of K_V subunits [36, 38–40], perhaps via engagement of different PKC isoforms [41]. This may contribute to selective control of K_V channel activity in response to different stimuli. In addition, increases in $[Ca^{2+}]_i$ have also been associated with inhibition of K_V channels [42, 43]. Considering that activation of G_qPCRs also increases $[Ca^{2+}]_i$, these data suggest that both PKC activity and elevated $[Ca^{2+}]_i$ could synergize to exacerbate K_V channel inhibition, which will result in membrane potential depolarization, activation of LTCCs, and vascular smooth muscle contraction.

In stark contrast to G_q signaling, activation of NO/PKG and G_s/PKA signaling is typically associated with stimulation of vascular K_V channels, including the members of the K_V1 and K_V7 subfamilies [1]. The functional consequence of this regulation is vascular smooth muscle relaxation. Phosphorylation of K_V channels by PKA is opposed by protein phosphatases such as PP2B, which will dephosphorylate the different subunits [44]. Whether G_sPCR/PKA/PP2B regulation of K_V channels requires scaffolding proteins that could target all components of the signaling complex within close proximity to the channels is not well understood. A recent series of studies have demonstrated that the scaffold protein postsynaptic density 95 (PSD95), which was thought to be a neuronal-specific protein, is expressed in vascular smooth muscle (**Figure 2**) [45, 46]. Intriguingly, PSD95 was found to be necessary for basal- and isoproterenol-induced PKA-mediated activation of K_V1.X channels that resulted in vascular smooth muscle relaxation [45–47]. This was due to the formation of a distinctive PSD95-mediated signaling complex involving the β1-adrenergic receptor (β1 AR)-, PKA-, and $K_V1.2$-containing channels [45, 47]. Since PSD95 is associated with AKAP5 in neurons [48], the argument was made that the PSD95-AKAP5 complex may be essential for PKA targeting and regulation of $K_V1.2$ function and that this will have an impact on vascular smooth muscle excitability and vascular reactivity [47]. However, additional studies have found that K^+ currents produced by K_V1.X and K_V2.X subunits and BK_{Ca} channels are of similar magnitude in wild-type and AKAP5-depleted ($AKAP5^{-/-}$) vascular smooth muscle cells [5, 49, 50]. These results suggest that, at least basally, AKAP5 is not necessary, and PSD95 may be sufficient for PKA-dependent regulation of K^+ channels in these cells. β-Adrenergic stimulation has also been found to regulate K_V7 channels leading to vasorelaxation [51], but whether a scaffold protein is mediating these effects is unknown and thus requires further examination. In addition, it is also unclear how K_V2-containing channels, which contribute about 70% of the K_V current in mouse cerebral and mesenteric vascular smooth muscle [49], are regulated by PKA signaling, presenting another area of further research.

3.2 BK_{Ca} channels

BK_{Ca} channels are abundantly expressed in vascular smooth muscle cells [52]. The pore-forming α subunit (BK_{Ca} α1.1) assembles into tetramers to form a functional channel, but unlike TRP and K_V subunits, it contains seven transmembrane domains and a heavily regulated long carboxyl terminal. Further regulation of BK_{Ca} channel function is conferred by the association of the α subunit with accessory β1 and LRRC26 γ subunits, which increases the Ca^{2+} sensitivity of the channel [52]. BK_{Ca} channels are sensitive to voltage and Ca^{2+}. The Ca^{2+} source to activate BK_{Ca} channels comes from the release of Ca^{2+} from the SR via RyRs (**Figure 2**) [20]. This complex does not depend on direct protein-protein interactions between BK_{Ca} channels and RyR but requires that both of these ion channels are in close physical proximity to each other. As mentioned above, data also have suggested that TRPV4 is responsible for triggering RyR activity [19], and thus, these channels have been proposed to form part of the same signaling complex with BK_{Ca} and RyRs in vascular smooth muscle. Intriguingly, LTCCs have been found to have an indirect or "loose" coupling with RyRs that could enable their activation [53], perhaps by modulating the SR Ca^{2+} concentration [54]. Additional components of the same or similar signaling complexes (see T-type Ca^{2+} channels section) with divergent roles have been identified [1, 7, 13, 55], but an integrated model remains to be defined. Regardless, the activation of these channels should result in K^+ efflux that hyperpolarizes the plasma membrane of vascular smooth muscle cells leading to relaxation. The physiological influence of BK_{Ca} channels in control of membrane potential and vascular tone, however, seems to be species-, stimulus-, and vessel-dependent, even when compared between different orders of the same vascular tree. Indeed, studies have reported a clear involvement of BK_{Ca} channels in tonic negative feedback regulation of membrane potential and pressure-induced constriction, even in human resistance arteries, whereas others have failed to establish a relationship (exemplary studies in [56–62]). The differences may be associated with disparities in the BK_{Ca} channel's Ca^{2+} sensitivity and/or subunit expression levels and composition in the different vascular beds. A recent and broader discussion on this topic can be found in [1]. Thus, given the distinctive role of BK_{Ca} channels in the vasculature, further research on how these channels are regulated to control vascular tone in different vascular beds during physiological and pathological conditions is warranted.

Generally, vasoactive agents that act via activation of the G_s/adenylyl cyclase (AC)/PKA and NO/soluble guanylyl cyclase (sGC)/PKG axes potentiate BK_{Ca} channel activity, whereas those acting through the G_q/phospholipase C (PLC)/PKC axis inhibit the channel [1, 52]. Vasoactive agents may also regulate BK_{Ca} channel activity indirectly by modulating the function of RyR in the SR or other ion channels (e.g., L-type $Ca_V1.2$ and T-type $Ca_V3.2$ channels) in the plasma membrane that are involved with direct activation of RyR or SR Ca^{2+} refilling [54, 63, 64]. Although there is some evidence that scaffold proteins such as AKAP5 may help target signaling molecules to BK_{Ca} channels [65], whether this indeed occurs in native vascular smooth muscle cells, as well as its functional relevance, is unclear. Considering the role of BK_{Ca} channels in negative feedback regulation of vascular tone, the physiological implications of vasoactive agents acting through distinct G_xPCRs on BK_{Ca} channels will be either vascular smooth muscle relaxation or contraction.

3.3 K_{ir} channels

As their name implies, K_{ir} channels produce an inward current. This current is observed at a potential negative to the K^+ equilibrium potential that helps stabilize the resting E_m [1, 66]. They also produce a small outward current at depolarizing potentials that serves as an electrical amplifier to magnify hyperpolarization. Inward rectification occurs due to voltage-dependent blockade of the channel by polyamines and Mg^{2+}. K_{ir} channels are regulated by lipids (e.g., phosphatidylinositol 4,5 bisphosphate (PIP_2) and cholesterol) [1, 66]. Intriguingly, a recent study found that cholesterol, but not PIP_2, regulates K_{ir} channel activity in the cerebral vascular smooth muscle [67], suggesting that these channels may be distinctively modulated by lipids depending on their tissue distribution. A functional K_{ir} channel is formed when four pore-forming α subunits, each containing two membrane-spanning domains, come together. Two main α subunits (e.g., $K_{ir}2.1$ and $K_{ir}2.2$) have been identified in vascular smooth muscle from multiple species [68–71]. Intriguingly, the expression of these subunits in a specific vascular bed may be species-dependent. Accordingly, although K_{ir} subunit expression and channel activity have been extensively reported in murine cerebral vascular smooth muscle [67–69, 71], minimal, if any, K_{ir} subunit expression and channel activity were found in the human cerebral vascular smooth muscle [72]. The functional implication of the activation of these channels in vascular smooth muscle is relaxation. K_{ir} channel activity can be modulated by vasoactive agents with those acting through the G_q/PLC/PKC axis, inducing channel inhibition, and those acting on the G_s/AC/PKA pathway, promoting channel activity [1]. The physiological relevance of these regulatory mechanisms on K_{ir} channels and their control of vascular function are less well understood and therefore are in need of further evaluation.

4. Voltage-gated Ca^{2+} channels

Vascular smooth muscle cells express several subtypes of VGCCs [9]. These channels have been shown to be important for vascular smooth muscle contraction, and some subtypes have been implicated in relaxation mechanisms. In this section, we will focus on the role of two key subtypes of VGCCs, namely, LTCCs and T-type Ca^{2+} channels (TTCCs), in regulation of vascular smooth muscle excitability.

4.1 L-type Ca^{2+} channel $Ca_V1.2$

The L-type Ca^{2+} channel $Ca_V1.2$ (i.e., LTCCs) is essential for vascular smooth muscle contraction and vascular reactivity. Therefore, they play a key role in controlling blood flow and blood pressure [33, 73]. LTCCs are comprised of a pore-forming α_{1c} subunit and auxiliary β, $\alpha_2\delta$, and γ subunits that modulate channel function and trafficking [74]. The α_{1c} subunit contains four homologous domains (I, II, III, IV). Each domain comprises of six membrane-spanning segments (S1–S6) with intracellular amino- and carboxyl termini, which contain many regions relevant for channel regulation and control of cell excitability. In vascular smooth muscle, expression of the α_{1c} subunit is critical for pressure-induced constriction as evidenced by an absence of myogenic response after LTCC blockade and depletion of the $Ca_V1.2$-α_{1c} subunit in mice (e.g., SMAKO mouse) [33, 73, 75]. The auxiliary subunits α2 and δ are the product of the same gene that gets proteolytically cleaved after translation but remains connected by disulfide bonds, which give rise to the mature subunit. The $\alpha_2\delta$ subunit has been linked to regulation of α_{1c} subunit surface expression that controls Ca^{2+} influx in vascular smooth muscle and the level of myogenic constriction [76]. The β subunit, which remains cytoplasmic, also contributes to the α_{1c} subunit surface expression and channel regulation and therefore can modulate vascular smooth muscle excitability in health and disease [77, 78]. Unlike the other subunits, the expression, regulation, and function of the γ subunit in vascular smooth muscle are unclear and likely the subject of further research.

LTCCs in vascular smooth muscle are distinctively regulated by the G_s/AC/PKA, NO/sGC/PKG, and G_q/PLC/PKC axes [9, 79]. Accordingly, the NO/sGC/PKG signaling axis has been shown to inhibit vascular LTCCs [80]. This has been associated with a reduction in $[Ca^{2+}]_i$ that may be part of the vasodilatory mechanism underlying the activation of this pathway [79]. Receptor-mediated signaling via the

G_q/PLC/PKC axis typically results in potentiation of LTCC activity [9, 79, 81]. The functional effects of this G_q/PLC/PKC-mediated activation of LTCCs are vascular smooth muscle contraction and an increase in vascular tone. Intriguingly, activation of the G_s/AC/PKA axis has been shown to inhibit, activate, or produce no effect on vascular LTCC activity (**Figure 3**) [79]. Irrespectively of this however, PKA signaling has been generally linked with vasodilation, thus raising questions about the functional relevance, if any, of this kinase in the regulation of vascular LTCCs. Intriguingly, recent studies revealed that elevations in extracellular D-glucose (HG) potentiate LTCC activity via a G_s/AC/PKA pathway in vascular smooth muscle [82–85]. This HG-induced PKA-dependent activation of LTCCs resulted in increased global $[Ca^{2+}]_i$ and vasoconstriction, thus providing the first example of a PKA-dependent pathway underlying vascular smooth muscle contraction. Future studies should further examine the in vivo relevance of this pathway.

LTCC regulation by G_s/AC/PKA and G_q/PLC/PKC axes in vascular smooth muscle is mediated by AKAP5 (**Figure 3**) [85, 86]. The involvement of the scaffold in this regulation was initially speculated from total internal reflection fluorescence (TIRF) microscopy experiments that optically recorded the activity of single or clusters of LTCCs [87, 88]. From these experiments, it was clear that the activity and location of functional LTCCs were heterogeneous throughout the surface membrane of vascular smooth muscle cells [81, 87, 89]. Whereas some LTCCs showed stochastic activity with low Ca^{2+} flux and duration of events, others had persistent activity characterized by increased Ca^{2+} flux and events with prolonged open time that were produced by the opening of two or more channels [81, 87, 89–91]. The

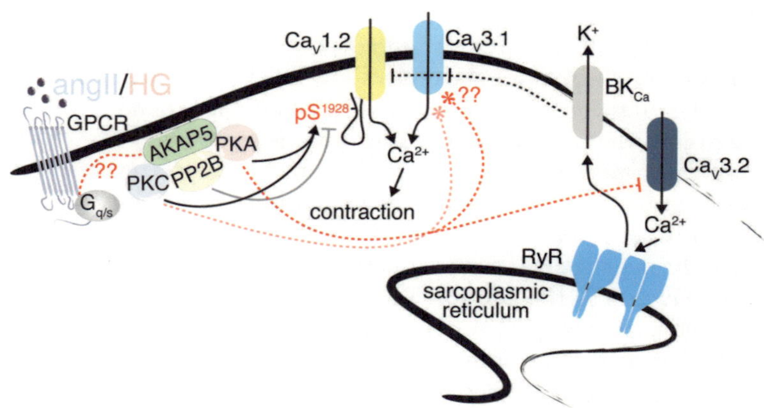

Figure 3.
Regulation of vascular smooth muscle excitability by voltage-gated Ca2+ channels. Ca^{2+} influx via L-type Ca^{2+} channel $Ca_V1.2$ is essential for vascular smooth muscle contraction [33]. Their activity is regulated by G_s/PKA and G_q/PKC signaling pathways upon activation of a specific GPCR by a given stimulus (e.g., angII or HG). The association of $Ca_V1.2$ with these signaling pathways is orchestrated by AKAP5 [86]. PKA (and perhaps PKC) augments L-type Ca^{2+} channel $Ca_V1.2$ activity by increasing $Ca_V1.2$ phosphorylation at serine 1928 (pS^{1928}) [82, 85]. The phosphatase PP2B suppresses enhancement of L-type Ca^{2+} channel $Ca_V1.2$ activity presumably by preventing/opposing channel phosphorylation in vascular smooth muscle cells (gray line). Two TTCC subtypes are expressed in vascular smooth muscle, with $Ca_V3.1$ contributing to contractile mechanisms and $Ca_V3.2$ forming a complex with RyR in the sarcoplasmic reticulum and BK_{Ca} channels in the surface membrane to foster relaxation [95, 100]. Although PKA has been shown to inhibit $Ca_V3.2$, whether this requires AKAP5 function is unclear (dotted red lines with perpendicular line at the end). It is also unclear whether AKAP5-anchored PKC and PKA regulate $Ca_V3.1$ channel activity (dotted light and dark red lines with star near $Ca_V3.1$). Finally, whether the GPCRs activated by angII and HG are targeted to specific complexes by AKAP5 is unclear (red dotted line with ?? symbols). The model was generated by taking in consideration studies cited and described above. GPCR = G-protein coupled receptors; angiotensin II = angII; high glucose = HG; AKAP5 = A kinase anchoring protein 5; protein kinase A= PKA; protein kinase C = PKC; protein phosphatase 2B; PP2B; phosphorylation at $Ca_V1.2$ serine 1928 = pS^{1928}; K^+ = potassium; Ca^{2+} = calcium; ryanodine receptors = RyR.

stochastic and persistent activity of LTCCs was modulated by membrane potential [92]. However, the occurrence of LTCCs with persistent activity is limited to specific regions of the surface membrane and has been demonstrated to be highly dependent on PKC activity and AKAP5 expression [81, 86]. The activity of phosphatases, such as PP2B, that are targeted to the channel by AKAP5, counteracts anchored kinase activity and restricts persistent LTCC activity (**Figure 3**) [89]. Accordingly, in vascular smooth muscle in which PKC is inhibited or cells from mice with genetically depleted PKC or AKAP5, the frequency of persistent LTCC activity is minimal [86, 87, 93]. In addition, PP2B inhibition stimulates persistent LTCC events in cells from wild type but not AKAP5$^{-/-}$ mice, suggesting that removing this "brake" facilitates kinase-mediated potentiation of channel activity [86, 89]. These results suggest an important role for AKAP5-anchored PKC and PP2B activity in modulating basal persistent LTCC activity. The physiological significance of these findings is underscored by data indicating that persistent LTCC events account for 50% of the total dihydropyridine-sensitive (e.g., LTCCs) Ca^{2+} influx at physiological membrane potentials [92], which is critical for vascular smooth muscle contractility in health and disease [82, 84, 86, 93].

4.2 T-type Ca^{2+} channels

T-type Ca^{2+} channels are formed by pore-forming α_1 subunits with similar topology as that of the LTCC α_{1c} subunit, but with no known auxiliary subunits that modulate channel function [74]. TTCCs are activated at more hyperpolarized potentials and show similar conductance with Ca^{2+} or Ba^{2+} as charge carriers. Vascular smooth muscle cells express several TTCC α_1 subunits, including $Ca_V3.1$ (α_{1G}) and $Ca_V3.2$ (α_{1H}) [9, 94–98]. Intriguingly, $Ca_V3.1$, which is found in murine vascular smooth muscle, seems to be replaced by $Ca_V3.3$ (α_{1I}) in human cells [96], suggesting that expression of TTCC α_1 subunits is species-dependent. TTCCs have been shown to contribute to vascular smooth muscle excitability in several vascular beds from different species [9, 98]. However, rigorous analysis revealed that different $Ca_V3.X$ subunits may have very divergent physiological responses. For instance, whereas $Ca_V3.1$ ($Ca_V3.3$) mediates low-pressure-induced constriction, $Ca_V3.2$ contributes to the negative feedback regulation of vascular tone by stimulating the RyR/BK_{Ca} axis (**Figure 3**) [64, 95, 96]. TTCCs can also be regulated by signaling molecules. Indeed, the NO/PKG and AC/PKA axes both inhibit vascular TTCCs [99, 100], which may have key implications in vascular smooth muscle excitability. Whether these signaling molecules are organized and targeted by scaffold proteins such as AKAPs to areas near TTCCs to fine-tune their function is unclear and therefore the subject of future studies.

5. Conclusions

Vascular smooth muscle excitability is exquisitely controlled by a repertoire of ion channels, which in themselves, are regulated by several vasoactive agents. The precise regulation of ion channels in vascular smooth muscle cells is essential for the dynamic adjustment of vascular tone necessary to maintain adequate tissue perfusion and blood pressure. Here, we have provided a brief overview of our current knowledge of key ion channels and their regulation by receptor-mediated signaling pathways that are activated by various vasoactive agents to modulate vascular smooth muscle excitability and therefore vascular tone. We focused on several TRP channels, multiple K^+ channel subtypes, and various classes of VGCCs. We emphasized ion channel regulation by signaling pathways associated with the G_s/AC/PKA,

NO/sGC/PKG, and G_q/PLC/PKC axes given their important role in modulating vascular smooth muscle excitability. When possible, we identified key gaps in knowledge, even in areas that have been extensively studied, which are fertile ground for further research. Because of the importance of all the ion channels and signaling pathways discussed above on vascular control, understanding how they are affected during pathological conditions is essential for the development of rational therapies to treat (micro)vascular diseases.

Acknowledgements

We thank the members of the Navedo Lab for critically reading early versions of the manuscript. This work was supported by NIH grants R01HL098200, R01HL121059 and R01HL149127 (to MFN), T32HL086350 (to AUS), and a UC Davis Academic Federation Innovative Development Award (to MN-C).

Conflict of interest

None.

Author details

Arsalan U. Syed, Thanhmai Le, Manuel F. Navedo and Madeline Nieves-Cintrón*
Department of Pharmacology, University of California Davis, Davis, CA, United States

*Address all correspondence to: mcnieves@ucdavis.edu

IntechOpen

© 2019 The Author(s). Licensee IntechOpen. This chapter is distributed under the terms of the Creative Commons Attribution License (http://creativecommons.org/licenses/by/3.0), which permits unrestricted use, distribution, and reproduction in any medium, provided the original work is properly cited.

References

[1] Tykocki NR, Boerman EM, Jackson WF. Smooth muscle ion channels and regulation of vascular tone in resistance arteries and arterioles. Comprehensive Physiology. 2017;7:485-581

[2] Bagher P, Segal SS. Regulation of blood flow in the microcirculation: Role of conducted vasodilation. Acta Physiologica. 2011;**202**:271-284

[3] Mederos YSM, Storch U, Gudermann T. Mechanosensitive Gq/11 protein-coupled receptors mediate myogenic vasoconstriction. Microcirculation. 2016;**23**:621-625

[4] Langeberg LK, Scott JD. Signalling scaffolds and local organization of cellular behaviour. Nature Reviews. Molecular Cell Biology. 2015;**16**:232-244

[5] Nieves-Cintron M, Syed AU, Nystoriak MA, Navedo MF. Regulation of voltage-gated potassium channels in vascular smooth muscle during hypertension and metabolic disorders. Microcirculation. 2018;**25**:e12423

[6] Amberg GC, Navedo MF. Calcium dynamics in vascular smooth muscle. Microcirculation. 2013;**20**:281-289

[7] Earley S, Brayden JE. Transient receptor potential channels in the vasculature. Physiological Reviews. 2015;**95**:645-690

[8] Narayanan D, Adebiyi A, Jaggar JH. Inositol trisphosphate receptors in smooth muscle cells. American Journal of Physiology. Heart and Circulatory Physiology. 2012;**302**:H2190-H2210

[9] Ghosh D, Syed AU, Prada MP, Nystoriak MA, Santana LF, Nieves-Cintron M, et al. Calcium channels in vascular smooth muscle. Advances in Pharmacology. 2017;**78**:49-87

[10] Nilius B, Szallasi A. Transient receptor potential channels as drug targets: From the science of basic research to the art of medicine. Pharmacological Reviews. 2014;**66**:676-814

[11] Reading SA, Earley S, Waldron BJ, Welsh DG, Brayden JE. TRPC3 mediates pyrimidine receptor-induced depolarization of cerebral arteries. American Journal of Physiology. Heart and Circulatory Physiology. 2005;**288**:H2055-H2061

[12] Xi Q, Adebiyi A, Zhao G, Chapman KE, Waters CM, Hassid A, et al. IP3 constricts cerebral arteries via IP3 receptor-mediated TRPC3 channel activation and independently of sarcoplasmic reticulum Ca2+ release. Circulation Research. 2008;**102**:1118-1126

[13] Adebiyi A, Zhao G, Narayanan D, Thomas-Gatewood CM, Bannister JP, Jaggar JH. Isoform-selective physical coupling of TRPC3 channels to IP3 receptors in smooth muscle cells regulates arterial contractility. Circulation Research. 2010;**106**:1603-1612

[14] Xu SZ, Boulay G, Flemming R, Beech DJ. E3-targeted anti-TRPC5 antibody inhibits store-operated calcium entry in freshly isolated pial arterioles. American Journal of Physiology. Heart and Circulatory Physiology. 2006;**291**:H2653-H2659

[15] Bulley S, Fernandez-Pena C, Hasan R, Leo MD, Muralidharan P, Mackay CE, et al. Arterial smooth muscle cell PKD2 (TRPP1) channels regulate systemic blood pressure. eLife. 2018;**7**:e42628

[16] Narayanan D, Bulley S, Leo MD, Burris SK, Gabrick KS, Boop FA, et al. Smooth muscle cell transient

receptor potential polycystin-2 (TRPP2) channels contribute to the myogenic response in cerebral arteries. The Journal of Physiology. 2013;**591**:5031-5046

[17] Welsh DG, Morielli AD, Nelson MT, Brayden JE. Transient receptor potential channels regulate myogenic tone of resistance arteries. Circulation Research. 2002;**90**:248-250

[18] Gonzales AL, Yang Y, Sullivan MN, Sanders L, Dabertrand F, Hill-Eubanks DC, et al. A PLC gamma1-dependent, force-sensitive signaling network in the myogenic constriction of cerebral arteries. Science Signaling. 2014;7:ra49

[19] Earley S, Heppner TJ, Nelson MT, Brayden JE. TRPV4 forms a novel Ca2+ signaling complex with ryanodine receptors and BKCa channels. Circulation Research. 2005;**97**:1270-1279

[20] Nelson MT, Cheng H, Rubart M, Santana LF, Bonev AD, Knot HJ, et al. Relaxation of arterial smooth muscle by calcium sparks. Science. 1995;**270**:633-637

[21] Mercado J, Baylie R, Navedo MF, Yuan C, Scott JD, Nelson MT, et al. Local control of TRPV4 channels by AKAP150-targeted PKC in arterial smooth muscle. The Journal of General Physiology. 2014;**143**:559-575

[22] Chen J, Crossland RF, Noorani MM, Marrelli SP. Inhibition of TRPC1/TRPC3 by PKG contributes to NO-mediated vasorelaxation. American Journal of Physiology. Heart and Circulatory Physiology. 2009;**297**:H417-H424

[23] Nishioka K, Nishida M, Ariyoshi M, Jian Z, Saiki S, Hirano M, et al. Cilostazol suppresses angiotensin II-induced vasoconstriction via protein kinase A-mediated phosphorylation of the transient receptor potential canonical 6 channel. Arteriosclerosis, Thrombosis, and Vascular Biology. 2011;**31**:2278-2286

[24] Albert AP, Large WA. Inhibitory regulation of constitutive transient receptor potential-like cation channels in rabbit ear artery myocytes. The Journal of Physiology. 2004;**560**:169-180

[25] Crnich R, Amberg GC, Leo MD, Gonzales AL, Tamkun MM, Jaggar JH, et al. Vasoconstriction resulting from dynamic membrane trafficking of TRPM4 in vascular smooth muscle cells. American Journal of Physiology. Cell Physiology. 2010;**299**:C682-C694

[26] Tajada S, Moreno CM, O'Dwyer S, Woods S, Sato D, Navedo MF, et al. Distance constraints on activation of TRPV4 channels by AKAP150-bound PKCalpha in arterial myocytes. The Journal of General Physiology. 2017;**149**:639-659

[27] Lopez-Lopez JR, Cidad P, Perez-Garcia MT. Kv channels and vascular smooth muscle cell proliferation. Microcirculation. 2018;**25**:e12427

[28] Perez-Garcia MT, Cidad P, Lopez-Lopez JR. The secret life of ion channels: Kv1.3 potassium channels and proliferation. American Journal of Physiology. Cell Physiology. 2018;**314**:C27-C42

[29] Dwenger MM, Ohanyan V, Navedo MF, Nystoriak MA. Coronary microvascular Kv1 channels as regulatory sensors of intracellular pyridine nucleotide redox potential. Microcirculation. 2018;**25**:e12426

[30] Zhong XZ, Abd-Elrahman KS, Liao CH, El-Yazbi AF, Walsh EJ, Walsh MP, et al. Stromatoxin-sensitive, heteromultimeric Kv2.1/Kv9.3 channels contribute to myogenic control of cerebral arterial diameter. J Physiol. 2010;**588**:4519-4537

[31] Jackson WF. Kv channels and the regulation of vascular smooth muscle tone. Microcirculation. 2018;**25**:e12421

[32] Knot HJ, Nelson MT. Regulation of membrane potential and diameter by voltage-dependent K^+ channels in rabbit myogenic cerebral arteries. The American Journal of Physiology. 1995;**269**:H348-H355

[33] Knot HJ, Nelson MT. Regulation of arterial diameter and wall [Ca^{2+}] in cerebral arteries of rat by membrane potential and intravascular pressure. The Journal of Physiology. 1998;**508** (Pt 1):199-209

[34] Dabertrand F, Kroigaard C, Bonev AD, Cognat E, Dalsgaard T, Domenga-Denier V, et al. Potassium channelopathy-like defect underlies early-stage cerebrovascular dysfunction in a genetic model of small vessel disease. Proceedings of the National Academy of Sciences of the United States of America. 2015;**112**:E796-E805

[35] Amberg GC, Rossow CF, Navedo MF, Santana LF. NFATc3 regulates Kv2.1 expression in arterial smooth muscle. The Journal of Biological Chemistry. 2004;**279**:47326-47334

[36] Clement-Chomienne O, Walsh MP, Cole WC. Angiotensin II activation of protein kinase C decreases delayed rectifier K+ current in rabbit vascular myocytes. The Journal of Physiology. 1996;**495**(Pt 3):689-700

[37] Mackie AR, Brueggemann LI, Henderson KK, Shiels AJ, Cribbs LL, Scrogin KE, et al. Vascular KCNQ potassium channels as novel targets for the control of mesenteric artery constriction by vasopressin, based on studies in single cells, pressurized arteries, and in vivo measurements of mesenteric vascular resistance. The Journal of Pharmacology and Experimental Therapeutics. 2008;**325**:475-483

[38] Kidd MW, Bulley S, Jaggar JH. Angiotensin II reduces the surface abundance of KV 1.5 channels in arterial myocytes to stimulate vasoconstriction. The Journal of Physiology. 2017;**595**:1607-1618

[39] Clement-Chomienne O, Walsh MP. Identification of protein kinase C isoenzymes in smooth muscle: Partial purification and characterization of chicken gizzard PKC zeta. Biochemistry and Cell Biology. 1996;**74**:51-65

[40] Ko EA, Park WS, Firth AL, Kim N, Yuan JX, Han J. Pathophysiology of voltage-gated K+ channels in vascular smooth muscle cells: Modulation by protein kinases. Progress in Biophysics and Molecular Biology. 2010;**103**:95-101

[41] Rainbow RD, Norman RI, Everitt DE, Brignell JL, Davies NW, Standen NB. Endothelin-I and angiotensin II inhibit arterial voltage-gated K+ channels through different protein kinase C isoenzymes. Cardiovascular Research. 2009;**83**:493-500

[42] Ishikawa T, Hume JR, Keef KD. Modulation of K+ and Ca2+ channels by histamine H1-receptor stimulation in rabbit coronary artery cells. The Journal of Physiology. 1993;**468**:379-400

[43] Gelband CH, Ishikawa T, Post JM, Keef KD, Hume JR. Intracellular divalent cations block smooth muscle K+ channels. Circulation Research. 1993;**73**:24-34

[44] Brignell JL, Perry MD, Nelson CP, Willets JM, Challiss RA, Davies NW. Steady-state modulation of voltage-gated K+ channels in rat arterial smooth muscle by cyclic AMP-dependent protein kinase and protein phosphatase 2B. PLoS One. 2015;**10**:e0121285

[45] Moore CL, Nelson PL, Parelkar NK, Rusch NJ, Rhee SW. Protein kinase A-phosphorylated KV1 channels in PSD95 signaling complex contribute to the resting membrane potential and diameter of cerebral arteries. Circulation Research. 2014;**114**:1258-1267

[46] Joseph BK, Thakali KM, Pathan AR, Kang E, Rusch NJ, Rhee SW. Postsynaptic density-95 scaffolding of shaker-type K(+) channels in smooth muscle cells regulates the diameter of cerebral arteries. The Journal of Physiology. 2011;**589**:5143-5152

[47] Moore CL, McClenahan SJ, Hanvey HM, Jang DS, Nelson PL, Joseph BK, et al. Beta1-adrenergic receptor-mediated dilation of rat cerebral artery requires shaker-type KV1 channels on PSD95 scaffold. Journal of Cerebral Blood Flow and Metabolism: Official Journal of the International Society of Cerebral Blood Flow and Metabolism. 2015;**35**:1537-1546

[48] Colledge M, Dean RA, Scott GK, Langeberg LK, Huganir RL, Scott JD. Targeting of PKA to glutamate receptors through a MAGUK-AKAP complex. Neuron. 2000;**27**:107-119

[49] Nieves-Cintron M, Nystoriak MA, Prada MP, Johnson K, Fayer W, Dell'Acqua ML, et al. Selective down regulation of Kv2.1 function contributes to enhanced arterial tone during diabetes. Journal of Biological Chemistry. 2015;**290**:7918-7929

[50] Nystoriak MA, Nieves-Cintron M, Nygren PJ, Hinke SA, Nichols CB, Chen CY, et al. AKAP150 contributes to enhanced vascular tone by facilitating large-conductance Ca2+−activated K+ channel remodeling in hyperglycemia and diabetes mellitus. Circulation Research. 2014;**114**:607-615

[51] Chadha PS, Zunke F, Zhu HL, Davis AJ, Jepps TA, Olesen SP, et al. Reduced KCNQ4-encoded voltage-dependent potassium channel activity underlies impaired beta-adrenoceptor-mediated relaxation of renal arteries in hypertension. Hypertension. 2012;**59**:877-884

[52] Latorre R, Castillo K, Carrasquel-Ursulaez W, Sepulveda RV, Gonzalez-Nilo F, Gonzalez C, et al. Molecular determinants of BK Channel functional diversity and functioning. Physiological Reviews. 2017;**97**:39-87

[53] Kotlikoff MI. Calcium-induced calcium release in smooth muscle: The case for loose coupling. Progress in Biophysics and Molecular Biology. 2003;**83**:171-191

[54] Takeda Y, Nystoriak MA, Nieves-Cintron M, Santana LF, Navedo MF. Relationship between Ca2+ sparklets and sarcoplasmic reticulum Ca2+ load and release in rat cerebral arterial smooth muscle. American Journal of Physiology. Heart and Circulatory Physiology. 2011;**301**:H2285-H2294

[55] Zhao G, Neeb ZP, Leo MD, Pachuau J, Adebiyi A, Ouyang K, et al. Type 1 IP3 receptors activate BKCa channels via local molecular coupling in arterial smooth muscle cells. The Journal of General Physiology. 2010;**136**:283-291

[56] Nelson MT, Brayden JE. Regulation of arterial tone by calcium-dependent K+ channels and ATP- sensitive K+ channels. Cardiovascular Drugs and Therapy. 1993;**7**(Suppl 3):605-610

[57] Dabertrand F, Nelson MT, Brayden JE. Acidosis dilates brain parenchymal arterioles by conversion of calcium waves to sparks to activate BK channels. Circulation Research. 2012;**110**:285-294

[58] Nieves-Cintron M, Syed AU, Buonarati OR, Rigor RR, Nystoriak MA, Ghosh D, et al. Impaired BKCa channel

function in native vascular smooth muscle from humans with type 2 diabetes. Scientific Reports. 2017;**7**:14058

[59] Khavandi K, Baylie RL, Sugden SA, Ahmed M, Csato V, Eaton P, et al. Pressure-induced oxidative activation of PKG enables vasoregulation by Ca2+ sparks and BK channels. Science Signaling. 2016;**9**:ra100

[60] Amberg GC, Bonev AD, Rossow CF, Nelson MT, Santana LF. Modulation of the molecular composition of large conductance, Ca^{2+} activated K^{+} channels in vascular smooth muscle during hypertension. The Journal of Clinical Investigation. 2003;**112**:717-724

[61] Jackson WF, Blair KL. Characterization and function of Ca(2+)-activated K+ channels in arteriolar muscle cells. The American Journal of Physiology. 1998;**274**:H27-H34

[62] Krishnamoorthy G, Sonkusare SK, Heppner TJ, Nelson MT. Opposing roles of smooth muscle BK channels and ryanodine receptors in the regulation of nerve-evoked constriction of mesenteric resistance arteries. American Journal of Physiology. Heart and Circulatory Physiology. 2014;**306**:H981-H988

[63] Jaggar JH, Porter VA, Lederer WJ, Nelson MT. Calcium sparks in smooth muscle. American Journal of Physiology. Cell Physiology. 2000;**278**:C235-C256

[64] Fan G, Kassmann M, Hashad AM, Welsh DG, Gollasch M. Differential targeting and signalling of voltage-gated T-type Cav 3.2 and L-type Cav 1.2 channels to ryanodine receptors in mesenteric arteries. The Journal of Physiology. 2018;**596**:4863-4877

[65] Liu G, Shi J, Yang L, Cao L, Park SM, Cui J, et al. Assembly of a Ca2+−dependent BK channel signaling complex by binding to beta2 adrenergic receptor. The EMBO Journal. 2004;**23**:2196-2205

[66] Longden TA, Nelson MT. Vascular inward rectifier K+ channels as external K+ sensors in the control of cerebral blood flow. Microcirculation. 2015;**22**:183-196

[67] Sancho M, Fabris S, Hald BO, Brett SE, Sandow SL, Poepping TL, et al. Membrane lipid-KIR2.x channel interactions enable hemodynamic sensing in cerebral arteries. Arteriosclerosis, Thrombosis, and Vascular Biology. 2019;**39**:1072-1087

[68] Sancho M, Samson NC, Hald BO, Hashad AM, Marrelli SP, Brett SE, et al. KIR channels tune electrical communication in cerebral arteries. Journal of Cerebral Blood Flow and Metabolism: Official Journal of the International Society of Cerebral Blood Flow and Metabolism. 2017;**37**:2171-2184

[69] Quayle JM, McCarron JG, Brayden JE, Nelson MT. Inward rectifier K+ currents in smooth muscle cells from rat resistance- sized cerebral arteries. The American Journal of Physiology. 1993;**265**:C1363-C1370

[70] Tajada S, Cidad P, Moreno-Dominguez A, Perez-Garcia MT, Lopez-Lopez JR. High blood pressure associates with the remodelling of inward rectifier K+ channels in mice mesenteric vascular smooth muscle cells. The Journal of Physiology. 2012;**590**:6075-6091

[71] Bradley KK, Jaggar JH, Bonev AD, Heppner TJ, Flynn ER, Nelson MT, et al. Kir2.1 encodes the inward rectifier potassium channel in rat arterial smooth muscle cells. The Journal of Physiology. 1999;**515**:639-651

[72] Sancho M, Gao Y, Hald BO, Yin H, Boulton M, Steven DA, et al. An assessment of KIR channel function

in human cerebral arteries. American Journal of Physiology. Heart and Circulatory Physiology. 2019;**316**:H794-H800

[73] Moosmang S, Schulla V, Welling A, Feil R, Feil S, Wegener JW, et al. Dominant role of smooth muscle L-type calcium channel Cav1.2 for blood pressure regulation. The EMBO Journal. 2003;**22**:6027-6034

[74] Catterall WA. Voltage-gated calcium channels. Cold Spring Harbor Perspectives in Biology. 2011;**3**:a003947

[75] Navedo MF, Amberg GC, Westenbroek RE, Sinnegger-Brauns MJ, Catterall WA, Striessnig J, et al. Ca(v)1.3 channels produce persistent calcium sparklets, but Ca(v)1.2 channels are responsible for sparklets in mouse arterial smooth muscle. Am J Physiol Heart Circ Physiol. 2007;**293**:H1359-H1370

[76] Bannister JP, Adebiyi A, Zhao G, Narayanan D, Thomas CM, Feng JY, et al. Smooth muscle cell alpha2delta-1 subunits are essential for vasoregulation by CaV1.2 channels. Circulation Research. 2009;**105**:948-955

[77] Kharade SV, Sonkusare SK, Srivastava AK, Thakali KM, Fletcher TW, Rhee SW, et al. The beta3 subunit contributes to vascular calcium channel upregulation and hypertension in angiotensin II-infused C57BL/6 mice. Hypertension. 2013;**61**:137-142

[78] Reimer D, Huber IG, Garcia ML, Haase H, Striessnig J. Beta subunit heterogeneity of L-type Ca(2+) channels in smooth muscle tissues. FEBS Letters. 2000;**467**:65-69

[79] Keef KD, Hume JR, Zhong J. Regulation of cardiac and smooth muscle Ca(2+) channels (Ca(V)1.2a,b) by protein kinases. Am J Physiol Cell Physiol. 2001;**281**:C1743-C1756

[80] Blatter LA, Wier WG. Nitric oxide decreases [Ca2+]i in vascular smooth muscle by inhibition of the calcium current. Cell Calcium. 1994;**15**:122-131

[81] Navedo MF, Amberg GC. Local regulation of L-type Ca(2)(+) channel sparklets in arterial smooth muscle. Microcirculation. 2013;**20**:290-298

[82] Prada MP, Syed AU, Buonarati OR, Reddy GR, Nystoriak MA, Ghosh D, et al. A Gs-coupled purinergic receptor boosts Ca(2+) influx and vascular contractility during diabetic hyperglycemia. eLife. 2019;**8**:e42214

[83] Syed AU, Reddy GR, Ghosh D, Prada MP, Nystoriak MA, Morotti S, et al. Adenylyl cyclase 5-generated cAMP controls cerebral vascular reactivity during diabetic hyperglycemia. J Clin Invest. 2019;**130**:3140-3152

[84] Navedo MF, Takeda Y, Nieves-Cintron M, Molkentin JD, Santana LF. Elevated Ca2+ sparklet activity during acute hyperglycemia and diabetes in cerebral arterial smooth muscle cells. American Journal of Physiology. Cell Physiology. 2010;**298**:C211-C220

[85] Nystoriak MA, Nieves-Cintron M, Patriarchi T, Buonarati OR, Prada MP, Morotti S, et al. Ser1928 phosphorylation by PKA stimulates the L-type Ca2+ channel CaV1.2 and vasoconstriction during acute hyperglycemia and diabetes. Science Signaling. 2017;**10**:eaaf9647

[86] Navedo MF, Nieves-Cintron M, Amberg GC, Yuan C, Votaw VS, Lederer WJ, et al. AKAP150 is required for stuttering persistent Ca2+ sparklets and angiotensin II-induced hypertension. Circulation Research. 2008;**102**:e1-e11

[87] Navedo MF, Amberg GC, Votaw VS, Santana LF. Constitutively active

L-type Ca^{2+} channels. Proceedings of the National Academy of Sciences of the United States of America. 2005;**102**:11112-11117

[88] Santana LF, Navedo MF. Molecular and biophysical mechanisms of Ca2+ sparklets in smooth muscle. Journal of Molecular and Cellular Cardiology. 2009;**47**:436-444

[89] Navedo MF, Amberg GC, Nieves M, Molkentin JD, Santana LF. Mechanisms underlying heterogeneous Ca2+ sparklet activity in arterial smooth muscle. The Journal of General Physiology. 2006;**127**:611-622

[90] Navedo MF, Santana LF. CaV1.2 sparklets in heart and vascular smooth muscle. Journal of Molecular and Cellular Cardiology. 2013;**58**:67-76

[91] Navedo MF, Cheng EP, Yuan C, Votaw S, Molkentin JD, Scott JD, et al. Increased coupled gating of L-type Ca2+ channels during hypertension and Timothy syndrome. Circulation Research. 2010;**106**:748-756

[92] Amberg GC, Navedo MF, Nieves-Cintron M, Molkentin JD, Santana LF. Calcium sparklets regulate local and global calcium in murine arterial smooth muscle. The Journal of Physiology. 2007;**579**:187-201

[93] Nieves-Cintron M, Amberg GC, Navedo MF, Molkentin JD, Santana LF. The control of Ca2+ influx and NFATc3 signaling in arterial smooth muscle during hypertension. Proceedings of the National Academy of Sciences of the United States of America. 2008;**105**:15623-15628

[94] Abd El-Rahman RR, Harraz OF, Brett SE, Anfinogenova Y, Mufti RE, Goldman D, et al. Identification of L- and T-type Ca2+ channels in rat cerebral arteries: Role in myogenic tone development. American Journal of Physiology. Heart and Circulatory Physiology. 2013;**304**:H58-H71

[95] Harraz OF, Abd El-Rahman RR, Bigdely-Shamloo K, Wilson SM, Brett SE, Romero M, et al. Ca(V)3.2 channels and the induction of negative feedback in cerebral arteries. Circulation Research. 2014;**115**:650-661

[96] Harraz OF, Visser F, Brett SE, Goldman D, Zechariah A, Hashad AM, et al. CaV1.2/CaV3.X channels mediate divergent vasomotor responses in human cerebral arteries. The Journal of General Physiology. 2015;**145**:405-418

[97] Harraz OF, Welsh DG. T-type Ca(2)(+) channels in cerebral arteries: Approaches, hypotheses, and speculation. Microcirculation. 2013;**20**:299-306

[98] Kuo IY, Wolfle SE, Hill CE. T-type calcium channels and vascular function: The new kid on the block? The Journal of Physiology. 2011;**589**:783-795

[99] Harraz OF, Brett SE, Welsh DG. Nitric oxide suppresses vascular voltage-gated T-type Ca2+ channels through cGMP/PKG signaling. American Journal of Physiology. Heart and Circulatory Physiology. 2014;**306**:H279-H285

[100] Harraz OF, Welsh DG. Protein kinase a regulation of T-type Ca2+ channels in rat cerebral arterial smooth muscle. Journal of Cell Science. 2013;**126**:2944-2954

Chapter 2

Cerebral Vascular Tone Regulation: Integration and Impact of Disease

Brayden Halvorson and Jefferson Frisbee

Abstract

This chapter summarizes the current knowledge regarding the regulation of the tone of cerebral resistance arteries under conditions of normal health and with the development of chronic diseases (e.g., metabolic disease). The work integrates the myogenic (pressure-induced) regulation of vascular tone, the impact of elevated luminal flow or shear stresses, that of local tissue metabolic activity on vascular tone and the concept of neurovascular coupling (linking neuronal activity to the impacts on vascular diameter). In addition, this work summarizes some of the recent work on how diseases such as type 2 diabetes impact the mechanisms of cerebrovascular tone regulation. It is anticipated that the current review will provide the reader with an up-to-date understanding of how the cerebral resistance vessels respond to changes in their local environment and contribute to the regulation of blood flow within the brain.

Keywords: cerebral, perfusion resistance, metabolic disease, vascular function, vascular disease risk

1. Introduction

The brain has a remarkably high metabolic rate and thus requires a highly disproportional amount of blood flow. Although its only 2% of body weight, the brain takes up 15–20% of cardiac output [1], making it one of the most highly perfused organs in the human body. This high metabolic rate coupled with its limited capacity for energy storage [2] necessitates heavy reliance on oxidative metabolism and thus requires constant blood flow to maintain nutrient and oxygen supply, remove waste products, and maintain a state of cerebral metabolic homeostasis. Severe underperfusion can quickly result in unconsciousness [3] and if prolonged, death [4]; while chronic mild under perfusion is associated with cognitive decline [5]. In addition to its high perfusion and metabolic rate, the cerebral circulation faces a unique challenge of being enclosed in the skull. This rigid structure prevents the expansion of tissue and extracellular fluid. Swelling within the skull from vasogenic edema leads to an increase in intracranial pressure which in turn can lead to neurologic complications or in more extreme cases death [1]. The unique challenges of the cerebral circulation, including intolerance to ischemia and edema, coupled with the paramount importance of maintaining constant nutrient and oxygen supply to cerebral tissue creates a need for precise regulation of cerebral blood flow and therefore the presence of redundant intrinsic mechanisms for its regulation. The anatomy of the brain vasculature ensures multiple routes for blood and oxygen delivery potentially allowing for perfusion even in cases of a blocked blood vessel [6]; however, acute regulation of flow is done primary by altering the diameter of blood vessels, and thus the resistance to flow. The major mechanisms of local regulation of vascular tone intrinsic to the cerebral vasculature include myogenic, shear, and metabolic based regulation. Although each mechanism has a discrete effect on vascular tone the integration of the different contributors to determine an appropriate level of tone is much more difficult to discern, especially in the cerebral circulation. These complex interactions allow for highly accurate control of cerebral blood flow in addition to protecting vulnerable downstream capillaries from high pressures and flow rates that could otherwise lead to edema; but, they also introduce several potential areas for failure. The intimate interactions of the various mechanisms of regulation of flow mean that the failure of one mechanism has the potential to initiate a cascade of events that results in inappropriate regulation of flow. As such abnormal execution of vascular tone regulation may form the basis of vascular pathologies [7].

One of these pathologies with a significant vascular component associated with impaired cerebral vascular tone regulation is metabolic syndrome (MetS). The incidence and prevalence of MetS is growing in Western society [8–10] and is contributing to decreased quality of life and increased economic burden. Thus an understanding of how it alters the cerebral circulation is crucial. MetS is categorized by a collection of metabolic risk factors including obesity, hypertension, atherogenic dyslipidemia and impaired glycemic control creating a pro-oxidant pro-inflammatory environment that raises the risk of developing impaired vascular structures and function [11–14]. These impairments are particularly detrimental when they affect the cerebral circulation and lead to cerebrovascular pathologies such as stroke or transient ischemic attack (TIA) due to the detrimental consequences associated with such events. However, cognitive impairments are not limited to individuals that have experienced an acute ischemic event since even in their absence MetS is strongly associated with impaired cognitive function and decreased quality of life [15–18]. Therefore preventing their occurrence by protecting the cerebrovasculature from functional and structural decline is paramount. This chapter will present a description of the local mechanisms involved in the regulation of cerebral vascular tone, how they integrate with one another and how they can be compromised in disease. Although impairments to the regulation of cerebral vascular tone are not limited to conditions associated with MetS this discussion will focus on the impact of MetS and its associated risk factors.

2. Myogenic mechanism

The myogenic mechanism which was first described by Bayliss is an intrinsic property of the vascular smooth muscle to respond to changes in intravascular pressure which is independent of other mechanisms of tone regulation including neural, metabolic, and hormonal influences [19]. The intrinsic nature of the myogenic response is supported by its existence in arteries and arterioles that have been sympathetically denervated and had their endothelium removed [20] thus leaving only the vessel itself to initiate and execute the response. The prototypical response of the vascular smooth muscle in response to an increase in intraluminal pressure is initial distension quickly followed by a constriction. The opposite can be said in situations of decreased intraluminal pressure; a fall in intraluminal pressure results in vessel collapse followed by dilation [21]. The myogenic response has several physiological roles including the establishment and maintenance of basal vascular tone (some degree of constriction) so that resistance may be increased or decreased by metabolic vasoconstrictors or vasodilators respectively, in order to regulate tissue perfusion. The establishment and maintenance of this partially constricted state in a pressurized vessel is referred to as myogenic tone. Additionally, the myogenic tone has a role in flow and pressure regulation. It functions by constricting to drop the pressure that reaches the downstream capillaries and protect them from edema or vascular remodeling associated with hypertension in the capillaries [22–25]. Equally important as protecting from hypertension is the ability of the vasculature to promote flow during low pressure by dilating. This ability to alter diameter over a range of pressures is referred to as myogenic reactivity. Beyond the local implications for the regulation of vascular tone, resistance to flow also has implications for systemic blood pressure since mean arterial pressure is the product of total peripheral resistance and cardiac output. This relationship illustrates system-wide implications of accurate vascular tone regulation. While the myogenic response is present throughout the body in a variety of vessels (arterioles, veins, lymphatic vessels) [26] the aforementioned functions of the myogenic response are particularly important in the cerebral circulation due to the catastrophic outcomes associated with under or over perfusion including unconsciousness and edema respectively. It is therefore not surprising that the most prominent myogenic response is found in the cerebral circulation with arterioles (resistance vessels) having the most pronounced response [1]. It should also be noted that the large arteries feeding the brain have a greater contribution to regulating vascular resistance in the cerebral circulation compared to other vascular beds, again providing evidence for the importance of regulating tone in the cerebral circulation.

A phase model of arterial myogenic behavior is commonly used to describe the response over a range of pressure. In the first phase, there is an initial development of myogenic tone at approximately 40–60 mmHg with increasing pressure up to that point causing passive distension. There is then a phase of myogenic reactivity in the pressure range of 60–140 mmHg and finally a phase of forced dilation at transmural pressures greater than approximately 140 mmHg [27].

2.1 Myogenic tone

Myogenic tone develops at approximately 40–60 mmHg and is characterized by an increase in intracellular Ca^{2+}, of about 200%, followed by a reduction in lumen diameter. This pressure causes cellular deformation, depolarization of the vascular smooth muscle cells (VSMC), and a significant increase in wall tension. Wall tension appears to be the controlled parameter in the myogenic response which is altered during an increase or decrease in transmural pressure. The myogenic response adjusts the diameter of the vessel in an attempt to restore or limit the change in basal wall tension through a negative feedback mechanism [28, 29]. The suggestion that wall tension is the controlled parameter is supported by its correlation with changes in cell calcium and myosin light-chain phosphorylation, a relationship that is not seen with vessel diameter [30, 31].

The mechanism of the myogenic response is attributed to stretch-activated ion channels: including L-type calcium channels, voltage-activated calcium channels and calcium activated potassium channels along with enzymatic mechanisms. Specifically, increased intraluminal pressure causes depolarization of the VSMC membrane and calcium influx by the opening of voltage-gated calcium channels (VGCCs), with the most prominent involvement being from $Ca_{V1.2}$. This influx of Ca^{2+} leads to increased myosin light-chain (MLC) phosphorylation which promotes increased actin/myosin interaction followed by cross-bridge cycling and cell shortening (vasoconstriction) [7, 24, 32–34]. The importance of Ca^{2+} influx in the generation of myogenic tone supported by its complete abolishment under Ca^{2+} free conditions [32, 35]; this is also a technique frequently used to study the passive mechanical characteristics of vessels to determine the degree of vascular remodeling since the VMSC exhibit a passive response (no force production) its strain is only due to the applied stress applied and the composition of the vessel [36]. Under physiological conditions the magnitude of the constrictor response to increased intraluminal pressure is limited by calcium-activated potassium channels that carry hyperpolarizing current proportional to the intracellular calcium concentration [35]. This negative feedback mechanism is supported by enhanced myogenic constriction being observed following blockade of calcium-activated potassium channel by specific inhibitors of these channels [37–39]. Additionally, at this pressure of 40–60 mmHg there is an activation of enzymatic systems and a complex interaction between matrix metalloproteins, the extracellular matrix, integrins and the cytoskeleton [40–42] that contribute to the myogenic reactivity at higher intraluminal pressures within the range of 60–140 mmHg [27]. This myogenic tone phase can also be characterized as the lower limit of autoregulation, which has important physiological implications. Below this pressure blood flow becomes dependent on blood pressure since the vessel cannot further dilate and begins to collapse as the pressure drops below this point [43]. Having an appropriate lower limit becomes especially important in situations of cerebral ischemia to allow restoration of flow in the presence of hypotension.

2.2 Myogenic reactivity

In this range of intraluminal pressure of 60–140 mmHg where the myogenic tone has already been established increases in pressure generally result in mild constriction and decreased pressure leads to mild dilation. Just as previously discussed for the generation of myogenic tone, increased pressure within this range leads to stretch, depolarization, and constriction of the vascular smooth muscle. However, in the myogenic reactivity phase, there is little change in vessel diameter across the range of pressures along with relatively small increases in Ca^{2+} (<20%) despite sizable increases in force production [27]. Multiple studies suggest an increased sensitivity to Ca^{2+} compared to the previous phase in the development of myogenic tone [20, 27, 44–48]. Increased sensitivity to Ca^{2+} is achieved by inhibition of myosin light chain phosphatase (MLCP) which promotes the accumulation of phosphorylated LC20 without an accompanying increase in calcium-induced myosin light chain kinase activity [49]. The presence of a contractile mechanism that does not require large variation of calcium, such as altering Ca^{2+} sensitivity requires less storage and transmembrane shuttling and is therefore advantageous in terms of conserving Ca^{2+} [35].

There are several proposed mechanisms that regulate Ca^{2+} sensitivity within this phase including, activation of protein kinase C (PKC), RhoA/Rho kinase pathways, and reactive oxygen species (ROS) [27, 35, 44, 46, 50]. The following studies provide evidence for the aforementioned mechanisms of enhanced Ca^{2+} sensitivity in this phase of the myogenic response through the utilization of specific inhibitors or transgenic animal models. Inhibition of PKC stops myogenic vasoconstriction in middle cerebral arteries with no impact on pressure-induced Ca^{2+} elevation [46]. Direct assessment of Ca^{2+} sensitivity by measuring the Ca^{2+}-tone relationship has consistency found decreased sensitivity during Rho kinase inhibition [20, 45]. ROK has also been reported to trigger smooth muscle depolarization during myogenic constriction and limit the extent of depolarization by opening delayed rectifier potassium channels [51, 52]. Arteries from transgenic animals missing NADPH oxidase function show an absence of myogenic activity [53], while mice deficient in superoxide dismutase, an endogenous antioxidant enzyme that catalyzes the breakdown of superoxide radical to H_2O_2, acquired enhanced myogenic reactivity [54]. Additional mechanisms that contribute to the myogenic reactivity phase independent of Ca^{2+} sensitivity include actin cytoskeleton reorganization and thin filament regulation [55, 56].

2.3 Forced dilation

Although the prototypical response of increased intraluminal pressure is a constriction, at excessively high pressures, beyond the autoregulatory range of approximately 140 mmHg, forced dilation often occurs [57]. This process results in a loss of myogenic tone, and thus results in an increase in vessel diameter, rapid increase in wall tension and significant elevation in Ca^{2+} (>50%) [27]. Although the name implies a degree of passiveness in the process, forced dilation is likely an active vasodilation involving K_{Ca} channels, nitric oxide (NO) and or ROS, which are activated to protect the arterial wall from damage [44, 57]. If the pressure is reduced to within the myogenic reactivity range reestablishment of tone and reduction in Ca^{2+} is observed.

3. Response to flow (shear stress)

An increase in flow leads to an increase in a frictional force known as shear that is detected by the endothelial cells lining the vessel lumen as blood moves through a vessel. As such, the magnitude of the shear force is proportional to blood flow. Shear catalyzes physiologically important responses in the cerebral vasculature such as encouraging reperfusion after ischemia, aiding in the hyperemic response to increased metabolic demand, and perhaps protection of downstream capillaries from edema and structural damage. Flow has been found to induce both constrictor and dilator pathways that act on the VSMC and result in a final level of tone taking into account the opposing processes. It is generally accepted that in the peripheral circulation flow leads to dilation; however, in the cerebral circulation the response is more controversial with both constriction and dilation being reported. These opposing observations may be because of a variety of factors including the area of the brain studied, the preparation used or because of interactions with other mechanisms affecting cerebral vascular tone. Within the cerebral circulation, the vertebrobasilar systems appear to elicit flow-mediated dilation, as measured in rats and mice [58] and humans [59]. The increase in flow is sensed by the endothelium, which initiates a negative feedback mechanism in an attempt to decrease the shear stress by dilating. Shear-induced dilation is largely endothelium-dependent and is at least partially mediated by NO [60]. The production of NO is controlled by the enzyme nitric oxide synthase (NOS), particularly endothelial NOS (eNOS), which catalyzes the formation of NO from L-arginine and is itself dependent on phosphorylation by Akt [61, 62]. The NO formed by eNOS then diffuses into the vascular smooth muscle where it activates soluble guanylyl cyclase (sGC) and increases cyclic guanine monophosphate (cGMP) and activating protein kinase G [63, 64]. Activation of protein kinase G opens large-conductance Ca^{2+}-activated K^{+} (BKCa) channels reducing intracellular Ca^{2+} which leads to relaxation of the vascular smooth muscle and thus vasodilation [65]. In addition to NO, eNOS can also lead to the formation of H_2O_2 which may also mediate flow-induced dilation [18]. During enzymatic cycling, eNOS produces oxygen radicals [66, 67] which, in the presence of sufficient antioxidants, are converted into H_2O_2 that may then, like NO, activate sGC and lead to dilation [68]. Other endothelium-dependent dilators such as prostaglandins do not appear to be involved in shear-induced dilation since the COX inhibitor indomethacin had no effect on the response to flow [60].

Constriction of cerebral vessels has also been reported in cats [69], rats [70] and human isolated cerebral arteries [71]. Constriction appears to predominate within the carotid circulatory area [58] especially when studies using *ex vivo* isolated vessels. In contrast to the dilator response, flow-induced constriction occurs independent of the endothelium [72, 73]. Although the mechanics for constriction in response to increased flow is not fully understood it appears to be mediated in part by a combination of integrin signaling, free radicals, and tyrosine kinase. Giving integrin-binding peptides, scavenging ROS by superoxide dismutase (SOD) or inhibiting tyrosine [69] all significantly reduced or eliminated constriction in response to flow. Just as in flow-induced dilation shear stress appears to be the initiating parameter in flow induced constriction whose force can be transduced by integrin signaling [69]. Additionally, Koller and Toth found that flow-induced constriction of cerebral vessels were blocked by inhibiting the synthesis of 20-HETE, thromboxane A_2 receptor, COX activity, and scavenging ROS. From that, they proposed that increase flow activates an AA cascade with metabolism by CYP450 4A enzymes resulting in the production of 20-HETE and ROS which contribute to the constriction via thromboxane A_2/prostaglandin H_2 receptors [71].

The differences in observed responses may also be dependent on whether the study was conducted *in vivo* or *ex vivo*. *Ex vivo* studies typically use a physiologic salt solution that has a much lower viscosity than blood and may be inadequate to stimulate the endothelium to produce endothelium-dependent dilation [74]. Consequently, the *ex vivo* responses to intraluminal shear have been shown to be endothelium-independent [69, 72]. Furthermore, *ex vivo* study generally lack pulsatile flow pumps but those that have been able to employ peristaltic or piston pump have indicated a role for shear-mediated regulation [68, 75, 76]. In contrast, in vivo study has produced more consistent shear-mediated dilation [62, 77]. Raignault et al. showed in a mouse model that the cerebrovascular endothelium optimally integrates shear stress to eNOS-mediated dilation under physiological pulse pressures, a phenomenon that was not seen in static flow conditions [76]. It appears that cerebral arteries are more responsive to a pulsatile environment with a viscose fluid (blood) as is experienced *in vivo* and consequently able to dilate in response to shear stress through endothelium-dependent production of NO [68, 76, 78].

Although flow induced constriction is not usually seen *in vivo* at high flow rates it may play a protective role in the cerebral circulation. Just as low flow situations are dangerous, especially in the brain, so too are instances of extreme flow; therefore it is important to have multiple regulatory mechanisms that are able to attenuate an increase in flow as a protective mechanism for the cerebral circulation from high volumes [1, 79, 80]. It is advantageous for the response to flow to adapt in accordance with other inputs on the regulation of vascular tone. This may be the reason why characterizing the response to flow in the cerebral circulation has been variable. One variable that may alter the response to flow is the absolute rate of flow. In isolated branches of MCA from rats vessels dilated with flow up to 10 μL/min but their diameter constricted back to baseline at higher levels of flow [81]. Shimoda et al. also found a biphasic flow-dependent response in anterior and middle cerebral arteries of neonatal pigs; flow rates between 0.077 and 0.212 mL/min constricted vessels while dilation occurred at higher flow rates up to 1.6 mL/min [82].

3.1 Flow and pressure

In a physiological setting multiple inputs are being processed by the cerebral vasculature leading to the generation of a certain level of tone. Pressure and shear stress exerted by flowing blood are two mechanical stimuli that have been described to play a major role in the regulation of vascular tone [83, 84]. It is therefore important to consider their interaction when determining the resultant effect on vascular tone. At high pressures (around >80 mmHg) cerebral vessels tend to constrict in response to flow [72, 85–87]. This biphasic response is further supported by findings from Garcia-Roldan and Bevan in isolated rabbit pial arterioles with flow rates from 0 to 20 μL/min at 90 mmHg but did not with the same flow at 60 mmHg [72].

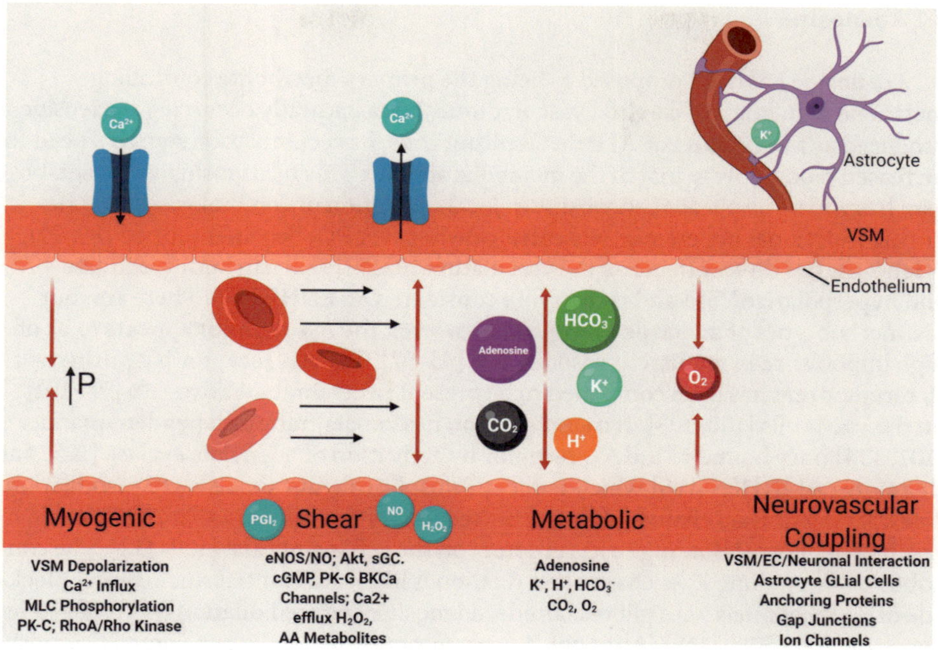

Figure 1.
A schematic representation of how intravascular pressure (myogenic), increased lumen flow rates (shear), tissue metabolism and neurovascular coupling can impact cerebral vascular tone regulation.

It is interesting that the tipping points for "high" and "low" pressures observed in previous studies are around 80 mmHg since, in resistance arterioles such as the MCA this pressure is commonly measured under physiological conditions. If 80 mmHg is indeed the point where higher pressures lead to constriction and lower to dilation in response to shear the resistance arterioles that are resting around this pressure would have the opportunity to tightly regulate the response to flow and shear thus allowing for precise control of cerebral blood flow. Please see **Figure 1** for a schematic representation of how flow and intravascular pressure can impact cerebral vascular tone.

4. Metabolic control

The mechanical stimuli of pressure and flow are generally thought to be important in setting the basal vascular tone so that metabolic influences are able to cause dilation or constriction depending on the needs of the cerebral tissue [24]. Metabolic control of vascular resistance is of particular importance in the cerebral circulation since cerebral tissue is extremely intolerant to ischemia [88]. As such, the cerebral circulation has a precise and highly localized coupling between the metabolic requirements of cerebral tissue and the magnitude of blood flow by controlling vascular resistance. There are numerous vasoactive metabolites that contribute to the control of cerebral vascular tone including adenosine, CO_2, H^+, O_2 and K^+. Increasing concentrations of adenosine, CO_2, H^+, and K^+ and decreased concentration of O_2 result in relaxation of VSM and dilation of cerebral resistance vessels. Each metabolite is associated with a cascade of events that ultimately either alters intracellular Ca^{2+} concentration or Ca^{2+} sensitivity of the VSM and results in a change in vessel diameter. Although the effects of each metabolite have been well characterized the relative importance of each along with its interaction with each other and other parameters of tone remains an area of further investigation. The discussion below is summarized in **Figure 1**.

4.1 Adenosine

Adenosine has been proposed as being the primary metabolite controlling metabolic regulation of cerebral vascular tone. It is a naturally occurring nucleoside produced as a byproduct of ATP metabolism; thus, its accumulation signals a need for increased blood flow to match the metabolic activity. This relationship with metabolism has widely implicated adenosine in local regulation of cerebral vascular tone during functional hyperemia, ischemia, or whenever PO_2 becomes limited [89–91]. Adenosine has direct effects on the vasculature [92, 93] that can both vasodilate and hyperpolarize VSM and is therefore considered an EDHF [94]. There are four distinct subtypes of adenosine receptors; however, the A_{2A} receptor appears to be of high importance in mediating vasodilation [95–97]. The A_{2A} receptor is a purinergic P_1 receptor that has been confirmed to be present in cerebral microvessels [98–101] on the VSM cells [102, 103]. It causes dilation in a concentration-dependent manner [101, 104] once bound to the A_{2A} receptor by activation of adenylate cyclase [105] and therefore cAMP [106–108] which reduces cytosolic calcium and leads to vasodilation. The opening of K_{ATP} channels also occurs secondary to the increase in cAMP levels as a result of adenosine binding to its receptors on the cell membrane [109–112]. The contribution of opening K_{ATP} channels to dilation is likely substantial since during blockade of K_{ATP} channels with glibenclamide, adenosine-induced dilation was reduced by approximately 50% [113]. Although A_{2A} receptors are generally considered the most important mediators of the effects of adenosine on vascular tone the A_{2B} receptors are also proposed to cause dilation through similar mechanisms as A_{2A} in addition to coupling to Gq proteins to produce Ca^{2+} mobilization by activation of phospholipase C and mitogen-activated protein kinase activation [98, 114, 115]. In addition to dilating the cerebral vasculature adenosine may also block vasoconstrictive signals in the parenchyma as evidenced by *in vitro* data from Gordon et al. [116]. When adenosine receptors are blocked with theophylline dilation was attenuated to arterial hypoxia [117]. Similarly the competitive adenosine receptor antagonist aminophylline causes a 20–30% decrease in CBF and cerebral oxygen delivery in normoxia [118].

4.2 PCO_2

Similar to many of the other metabolites discussed CO_2 tends to increase under conditions of increased metabolism without adequate flow to eliminate it from the area of production and thus its accumulation leads to dilation of the vasculature. High sensitivity to PCO_2 is unique to the cerebral circulation [119] causing approximately 3–6% increase and 1–3% decrease in flow per mmHg change in $PaCO_2$ above or below eupnoeic $PaCO_2$ respectively. This high sensitivity is seen throughout the arterial side of the vascular network including the large arteries in the neck [120] and large intracranial arteries [121–123] to the smallest pial arterioles [124] and parenchymal vessels [125–128]. There are likely several redundant mechanistic contributors to the sensitivity of the vasculature to PCO_2 which may contribute to the debate as to whether the dilation is triggered by increased PCO_2 or rather the accompanying increase H^+ concentration from the carbonic anhydrase reaction. There is strong support that the change in PCO_2 mediates at least in part alterations in cerebral vascular tone locally by changes in perivascular pH [107, 129–131] as evidenced by acidic and alkaline perfusate administered through an intracranial window. Experiments that have been able to alter pH and PCO_2 independently have provided evidence of the dependence of altered pH to initiate a change in cerebral vascular tone [74]. In a cat pial arteriole cranial window preparation, lowering pH along with hypercapnia resulted in no difference in the magnitude of vasodilation when compared to acidic isocapnia [132]. Interestingly, vessel tone was unaltered in response to intraluminal CO_2 change, suggesting that a change in superfusate pH is necessary to evoke a change in cerebral vascular tone [107, 133]. This is supported by the findings of unchanged CBF in humans [107] and animals [133] in response to changes in arterial pH, and highly localized pial arteriolar diameter changes in response to the application of acidic/basic solution into the perivascular space [131]. Additionally, when pH is maintained as seen in an experiment with artificial CSF pretreated with sodium bicarbonate, pial vessel dilation is eliminated in response to intraluminal hypercapnia [107, 134] further supporting that a change in superfusate pH is necessary to alter the cerebral vascular tone. In addition to its link with CO_2 and H^+ through the carbonic anhydrase reaction, there is some evidence that bicarbonate ion may independently influence vascular tone. Having the ability to sense and react to multiple parameters associated with acidosis allows for control of tone in response to not only the pH but also the cause of the disturbance (accumulation of CO_2 for instance). In isolated rat basilar arteries reduced $[HCO_3^-]_o$ has been shown to directly increase cerebral vascular tone (with pH maintained at 7.4 and CO_2 kept constant at 5%) through the binding of receptor protein tyrosine phosphatase through an endothelium-dependent response [135]. HCO_3^- may, therefore, stimulate soluble adenylate cyclase activity through a pH-independent mechanism [136]. If this is correct then the modest increase in $[HCO_3^-]$ during hypercapnia may aid in the dilation response to CO_2 but the opposing vasocontractile response to decreases in $[HCO_3^-]_o$ may limits the vasorelaxation caused by a reduction in pH [135]. Limiting vasorelaxation during acidosis is important to lessen the increase in capillary pressure associated with vasorelaxation of upstream arterioles. Hyper-relaxation may overload the capillaries leading to edema and damage thereby worsening the consequences of local inadequate perfusion.

4.3 PO_2

The effects of oxygen are unique in that its availability is required for aerobic metabolism rather than a byproduct like some of the metabolic factors discussed. Therefore it is not surprising that its abundance leads to vasoconstriction and its relative shortage leads to dilation. Although its availability is tightly linked to that of PCO_2 and H^+ and other metabolic byproducts in a physiological setting, studies have been able to discern its independent effect in the presence of otherwise constant conditions. Data from isolated arteries/arterioles suggest there is in fact an oxygen sensor independent of other vasoactive metabolic byproducts within the vascular wall itself [137–142]. Once the change in oxygen is sensed there are various mediators of hypoxic dilation including endothelial-derived NO [138, 141], prostanoids [137, 140, 141, 143], 20-HETE [141] and EDHF [139]; however, the contribution of each factor appears to be dependent on the severity of hypoxia [141]. In skeletal muscle, dilation from mild hypoxia (15% O_2) was mostly NO-dependent, while moderate (10% O_2) was mediated by a combination of increased PGI_2 and decreased 20-HETE, and severe (0% O_2) was almost entirely accounted for by an increase in PGI_2 [141]. In all cases, there appears to be significant involvement by the endothelium to mediate the dilation, which is further supported by a reduction in hypoxic dilation when isolated vessels were exposed to indomethacin (an inhibitor of AA metabolism and thus the production of PGI_2) [137, 144, 145] and to a lesser degree by L-NAME (an inhibitor of NO production from NOS) [145]. Therefore, PGI_2 is likely a substantial contributor to hypoxic dilation with a lesser but likely still significant role for NO. Human studies measuring CBF with pcMRI, instead of isolated vessel diameter as in the aforementioned studies, with and without administration of L-NAME suggest that hypoxic dilation is highly dependent on NO, with no change from baseline observed in hypoxia when L-NAME was administered [146]. These differences may be because of a species-specific response or differences in *ex vivo* and *in vivo* conditions.

Additional mechanisms of hypoxic dilation include adenosine whose action has previously been discussed. In hyperoxic conditions constriction is favored, mediated by a greater conversion of AA to 20-HETE as opposed to dilators such as PGI_2 [147].

4.4 Potassium and neurovascular coupling

Although K^+ is not directly a byproduct of a metabolite pathway it tends to increase in concentration when the frequency of neuronal depolarization is increased, and is therefore indicative of increased metabolism. K^+ channels are present in cerebrovascular smooth muscle cells and are important regulators of tone because of their ability to alter membrane potential [148, 149]. Although K^+ channels are present through the peripheral vasculature, the cerebral circulation has a unique anatomical feature that allows for intimate interaction of astrocytic endfeet and cerebral vessels. This tight interaction between the astrocytic endfeet and vasculature allows for precise localized changes in blood flow to match the site-specific neural activity in the brain [150, 151]. K^+ can therefore be thought of as a direct link between neuronal activity and blood flow. This pairing of neuronal metabolism with appropriate blood flow and is termed neurovascular coupling (NVC). NVC forms the mechanistic basis for neuroimaging techniques that are able to map changes in neuronal activity based on vascular responses such as changes in blood flow or oxygen saturation [152]. As such, an understanding of NVC is not only crucial to understand the regulation of cerebral vascular tone but is also needed to interpret these neuroimaging techniques including functional magnetic resonance imaging (fMRI), positron emission tomography (PET), and near-infrared spectroscopy [153].

NVC is dependent on the interactions of the neurovascular unit which is made up of three major components: the vascular smooth muscle, neuron, and astrocyte glial cell. Somewhat surprisingly, the initiation site of neurovascular coupling is at the level of capillaries which then leads to changes upstream; however, once the close anatomical locations of neurons to the capillaries (8–20 μm) are considered this becomes a logical point of initiation [154]. Not only does NVC rely on interactions apart from the vasculature but it also requires several structural components to facilitate cell-to-cell interactions of these different cell types. These components include gap junctions [155, 156], anchoring proteins [157] and specialized ion channels [158] expressed on cell–cell interface membranes. Neurons initiate NVC by generating direct signals that act on the vasculature and indirect signals that are transmitted through astrocytes and lead to increases in intracellular Ca^{2+} within the astrocyte. This is achieved by the glutamatergic synaptic activity initiating post-synaptic N-methyl-D-receptors (NMDA) and a-amino-3-hydroxy-5-methyl-4-isoxazolepropionic acid (AMPA) receptors. The increase in intracellular Ca^{2+} activates Ca^{2+}-dependent enzymes to produce vasodilators. Some of the enzymes activated include neuronal NOS (nNOS) and cyclooxygenase 2 (COX-2) [159, 160]. Glutamate also acts on metabotropic glutamate receptors in astrocytes increasing Ca^{2+} in these cells and leading to the production of vasoactive metabolites including adenosine, ATP, and K^+ that act of the VSMCs [161].

Astrocytes are anatomically well positioned to transmit signals from neurons to the vasculature because of their close proximity to the capillaries; however, in a vascular network both the downstream and upstream vasculature must work in concert in order to effectively regulate blood flow delivery. It appears that endothelial cells provide the crucial role of retrograde propagation of the vasomotor response allowing for coordination between up and downstream vasculature. Endothelial cells can produce numerous metabolites including NO, prostanoids, and endothelin to alter VSMCs constriction. Endothelial intermediate K^+ channels (K_{IR}) channels and small conductance K^+ channels (K_{SK}) have been implicated as a mechanistic contributor to the propagation of the vasodilatory signal upstream for a synchronized hemodynamic response between the capillaries and arterioles via myoendothelial gap junctions [162]. Secondary to the movement of hyperpolarizing current through gap junctions, K^+ efflux from K_{IR} and K_{SK} channels generates what has been described as a "K^+ cloud" [163] that also hyperpolarizes neighboring VSMC through both the inducible Na^+-K^+ electrogenic pump and stimulation of inward rectifying potassium channels in the VSMC of cerebral arteries and arterioles. In both cases, there is a net efflux of K^+ from VSMC [164–167] in conjunction with closed voltage-gated calcium channels and therefore results in the relaxation of the VSMC. The importance of K^+ channels in dilating the cerebral vasculature is supported by reduced K^+ conductance, depolarization and ultimately constriction seen in experimental inhibition of the channels [163]. For example, 6–10 mM K^+ applied to capillaries generated a hyperpolarizing response in endothelial cells which was transmitted upstream to penetrating arterioles hyperpolarizing and relaxing VSMCs [168]. This propagation of the vasodilation signal was blocked with inhibition of K_{IR} with barium or endothelial deletion of $K_{IR1.2}$ channels, supporting its role transmitting the hyperpolarizing current. Interestingly the capillaries themselves did not dilate to K^+ suggesting they act as a sensor detecting the signal for the upstream arterioles to be the effector of the response [161].

Although in a physiological setting K^+ tends to lead to vasodilation, the response to K^+ may be dependent on the $[K^+]$. In isolated cerebral arterioles low (<7 mM) and moderate (8–15 mM) increases in $[K^+]$ result in endothelium-independent dilations and sustained dilation respectively through mechanisms previously discussed [165]. Higher K^+ concentrations cause constriction; however, under normal physiological conditions this concentration is not reached and is only seen under pathological processes such as spreading depression and stroke [169, 170]. This response seen in isolated vessel has also been confirmed in vivo [151]. There is a combination of mechanisms at work in response to K^+ that may explain the opposing responses to different concentrations. Alone an increase in $[K^+]_o$ actually tends to depolarize neighboring VSMC resulting in constriction; this response predominates at high $[K^+]$. At low concentrations, the minor increase in $[K^+]_o$ would by itself lead to constriction but it also stimulates both K_{IR} channels [150, 171–174] and Na^+/K^+ ATPase [175–177] promoting K^+ efflux as previously described. This description of the electrophysiology at work in the VSMC is supported by E_m measurement in VSMCs when a solution containing $[K^+]$ (<20 mM) is applied and produces VSM hyperpolarization, whereas at higher K^+ concentrations depolarization predominates [166, 172, 178]. Ultimately the signals generated by neurons, astrocytes, and endothelial cells must be received and integrated into a final level of tone by the VSMCs. A schematic representation of the impact of neurovascular coupling on cerebral vascular tone is presented in **Figure 1**.

5. Effect of diseased states

Each component implicated in the regulation of tone has a multitude of signals being produce that is intended to affect a vasomotor response from the VSMCs. Signals from myogenic, shear, metabolic, and neurovascular influences may be additive or opposing in their effect on VSMCs and further combine with one another creating intricate and precise regulation of cerebral vascular tone. It is this intricacy however, that also introduces many possible steps in the pathway for an abnormal response to occur. As such changes to the regulation of tone may form the basis of several pathologies including MetS. The contributing risk factors associated with MetS alter the local regulation of cerebral vascular tone by inducing changes in both the structure and function of the vessels. These risk factors include hypertension, T2DM, and obesity which promote a pro-inflammatory pro-oxidant state. Functionally, MetS is highly linked to increased smooth muscle activation and endothelial dysfunction which has important implications for the ability of a vessel to dilate in response to a multitude of stimuli previously discussed including hypotension, shear, hypoxia, and other metabolic stimuli. Increased myogenic properties are consistently overserved in MetS in multiple vascular beds [179, 180] including in the cerebral circulation [181–183]. This increase in constriction may be due to both a decrease in buffering capacity from endothelial dysfunction and alterations in the vascular smooth muscle itself [183, 184]. Endothelial dysfunction also has implication for shear-induced dilation since it is highly dependent on NO bioavailability which has consistently been shown to be reduced in MetS. The chronic inflammatory state seen in MetS likely contributes to the reduction in NO bioavailability due to increased scavenging to the produced NO by reactive oxygen species since this reduction in NO bioavailability is consistently reported to evolve in parallel with oxidant stress and the development of a chronic inflammatory state [185, 186]. This is supported by improved dilatory reactivity of MCA with the pretreatment of the cell-permeable superoxide dismutase mimetic TEMPOL in a model of T2DM [145]. Interestingly some studies have actually shown an increase in eNOS expression which may be an attempt to compensate for the increased scavenging; however, they too continue to find reduced dilator reactivity [187].

Aside from reduced NO bioavailability, a shift in arachidonic acid metabolism toward constrictors and away from dilators that are highly responsible for hypoxic dilation has been demonstrated [145]. Thus not only does endothelial dysfunction seen in MetS increase cerebrovascular resistance by decreased dilator metabolite production it may also promote the production of constrictors that exacerbate the impaired dilation capacity of cerebral vessels [184, 188]. A change in sensitivity to various metabolites in addition to their differential production may also contribute to differential vasomotor responses. For example, studies using SNP, an exogenous NO donor, while blocking endogenous NO production by eNOS using L-NAME, found smaller relaxation of the MCA in spontaneously hypertensive rats which was attributed to a decreased expression of soluble guanylate cyclase [189–192]. In a model of T2DM decreased sensitivity of MCA to the PGI_2 analog iloprost was also found suggesting that both decrease production and sensitivity of dilators may be contributing to the impaired dilation of the cerebral circulation in MetS [145].

Impaired dilation in response to exogenous dilator metabolites may also be due to vascular remodeling. Structural changes are an important consideration since even if the smooth muscle of a cerebral vessel is able to relax due to metabolic influences resulting in hyperpolarization, remodeling may prevent an increase in lumen diameter which is ultimately the major contributor to acute changes in resistance and thus the regulator of flow. Hypertension is largely implicated in the thickening of the vascular smooth muscle as well as increasing the ratio of collagen to elastin in the vessel. High intraluminal pressure increases the shear stress exerted on the vascular endothelium which normally could be restored to baseline by NO-induced vasodilation [193–195]; however, in a disease state with impaired NO production, there is a reduced ability to dilate resulting in endothelial damage and upregulation of atherogenic genes [193–195]. As a means of protection from chronic increased shear stress and wall tension that may lead to downstream edema cerebral vessels tend to hypertrophy with chronic hypertension, but this protective hypertrophy is also detrimental [188]. Since wall tension is equal to intraluminal pressure X radius and wall stress is wall tension/wall thickness, hypertension-induced hypertrophy and inward remodeling resulting in a decrease in radius and increase wall thickness can normalize both the wall tension and wall stress [188]. Although this remodeling may be protective in regards to increases in pressures and protecting the downstream capillaries from edema it increases the cerebrovascular resistance and limits the dilation reserve during hypotension and therefore presents itself as a right-shift in the autoregulatory range [196, 197]. The development of myogenic tone at higher pressure implies an increased lower limit of autoregulation. This predisposes cerebral tissue to reduced blood flow during hypotension. When pressure drops below the lower limit flow becomes dependent on the passive diameter of the vessel. Not only does an impaired lower limit predispose hypertensive individuals to ischemia but the reduced passive diameter from vascular remodeling further compromises flow to cerebral tissue resulting in hypoxic areas [43].

In addition to hypertrophy and inward remodeling, there is substantial arterial stiffening commonly seen in MetS. The pro-oxidant stress of ROS may interact with components of the perivascular matrix and initiate collagen cross-linking and deposition as well as the breakdown of elastin making the vessel less distensible [198]. This is measured by a left shift in the stress-strain curve of isolated cerebral vessels under passive conditions achieved by using a Ca^{2+} free solution preventing the development of tone [183]. The stiffening vessel from increased collagen to elastin ratio is made worse by the thickening of vessel walls previously discussed. In a model of T2DM with hypertension significant collagen deposition in addition to medial hypertrophy increasing the wall the lumen ratio and stiffness of the rats MCA [199] was demonstrated while T2DM in the absence of hypertension does not appear to induce structural changes to the cerebral vasculature [145]. This along with data that suggests the increase in arterial stiffness seems to follow a time course similar to that of the onset of hypertension suggests a strong relationship between hypertension and vascular remodeling [200]. Chronic uncontrolled hyperglycemia and inflammation do tend to lead to the development of hypertension and thus contribute to changes in the composition of cerebral vessels and it is likely the combination of both seen in MetS increases the degree to which remodeling occurs. A summary of the impact of disease states on the regulation of cerebral vascular tone is presented in **Figure 2**.

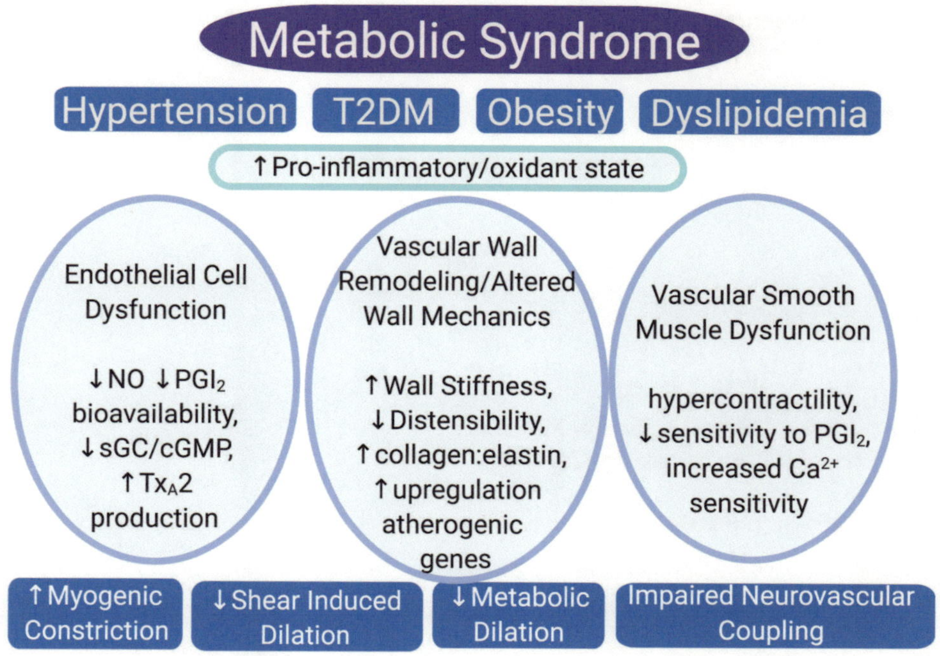

Figure 2.
A schematic representation of how the presence of metabolic syndrome and the major constituent pathologies can impact the integrated regulation of cerebral vascular tone.

6. Conclusions

In conclusion, the regulation of vascular tone involves a complex set of pathways with myogenic, shear, and metabolic control. The mechanical influences of pressure and flow serve as a stimulus for the myogenic and shear responses to set a basal level of tone over a wide range so that metabolic factors have room to produce vasoactive responses on the vasculature. Due to the paramount importance of precise cerebral blood flow control these mechanisms are particularly pronounced and redundant in the cerebral circulation allowing for greater protection against insufficient perfusion or edema and capillary damage in situations of hypotension and hypertension respectively. However, due to the complexity of these homeostatic blood flow mechanisms there is the potential for the development of a pathological state. MetS presents a constellation of cardiovascular risk factors that are highly linked to the development of such cerebrovascular pathologies increasing the risk of stroke, TIA, and vascular dementia. The risk factors associated with MetS result in vascular remodeling which decreases the lumen size and increases stiffness and when paired with endothelial dysfunction and increased activation of the vascular smooth muscle it promotes increased cerebrovascular resistance. This right shifts the autoregulatory zone of myogenic regulation allowing for enhanced protection from hypertension but leaves cerebral tissue vulnerable to underperfusion. Controlling these risk factors and well as implementing targeted therapeutic strategies aimed at ameliorating the regulation of cerebrovascular tone has the potential to restore function in the cerebral circulation and improve current negative outcomes associated with MetS and cerebrovascular dysfunction.

Acknowledgements

This work was supported by Canadian Institutes of Health Research #389769 and Natural Sciences and Engineering Research Council (Canada) RGPIN-2018-05450. All figures were created with Biorender.com.

Author details

Brayden Halvorson and Jefferson Frisbee*
University of Western Ontario, Canada

*Address all correspondence to: jfrisbee@uwo.ca

IntechOpen

© 2020 The Author(s). Licensee IntechOpen. This chapter is distributed under the terms of the Creative Commons Attribution License (http://creativecommons.org/licenses/by/3.0), which permits unrestricted use, distribution, and reproduction in any medium, provided the original work is properly cited.

References

[1] Cipolla MJ. The Cerebral Circulation. San Rafael, CA, USA: Morgan & Claypool Life Sciences; 2010

[2] Brown AM, Ransom BR. Astrocyte glycogen and brain energy metabolism. Glia. 2007;**55**(12):1263-1271

[3] Van Lieshout JJ, Wieling W, Karemaker JM, Secher NH. Syncope, cerebral perfusion, and oxygenation. Journal of Applied Physiology. 2003;**94**:833-848

[4] Smith BA, Clayton EW, Robertson D. Experimental arrest of cerebral blood flow in human subjects: The Red Wing studies revisited. Perspectives in Biology and Medicine. 2011;**54**(2):121-131

[5] Ng TP, Feng L, Nyunt MSZ, Feng L, Gao Q, Lim ML, et al. Metabolic syndrome and the risk of mild cognitive impairment and progression to dementia follow-up of the Singapore longitudinal ageing study cohort. JAMA Neurology. 2016;**73**(4):456-463

[6] Gebremedhin D, Gopalakrishnan S, Harder DR. Endogenous events modulating myogenic regulation of cerebrovascular function. Current Vascular Pharmacology. 2014;**12**(6):810-817

[7] El-Yazbi AF, Abd-Elrahman KS. ROK and arteriolar myogenic tone generation: Molecular evidence in health and disease. Frontiers in Pharmacology. 2017;**8**(87):1-10

[8] Boutayeb A, Boutayeb S, Boutayeb W. Multi-morbidity of non communicable diseases and equity in WHO Eastern Mediterranean countries. International Journal for Equity in Health. 2013;**12**(1):60

[9] De Carvalho VF, Bressan J, Babio N, Salas-Salvadó J. Prevalence of metabolic syndrome in Brazilian adults: A systematic review. BMC Public Health. 2013;**13**:1198

[10] Fu J, Prasad HC. Changing epidemiology of metabolic syndrome and type 2 diabetes in Chinese youth. Current Diabetes Reports. 2014;**14**:447

[11] DeVallance E, Fournier SB, Donley DA, Bonner DE, Lee K, Frisbee JC, et al. Is obesity predictive of cardiovascular dysfunction independent of cardiovascular risk factors? International Journal of Obesity. 2015;**39**:244-253

[12] Donley DA, Fournier SB, Reger BL, DeVallance E, Bonner DE, Olfert IM, et al. Aerobic exercise training reduces arterial stiffness in metabolic syndrome. Journal of Applied Physiology. 2014;**116**(11):1396-1404

[13] Ferdinand KC, Rodriguez F, Nasser SA, Caballero AE, Puckrein GA, Zangeneh F, et al. Cardiorenal metabolic syndrome and cardiometabolic risks in minority populations. Cardiorenal Medicine. 2014;**4**:1-11

[14] Shin JA, Lee JH, Lim SY, Ha HS, Kwon HS, Park YM, et al. Metabolic syndrome as a predictor of type 2 diabetes, and its clinical interpretations and usefulness. Journal of Diabetes Investigation. 2013;**4**(4):334-343

[15] Rosenberg GA. Inflammation and white matter damage in vascular cognitive impairment. Stroke. 2009;**40**(3):S20-S23

[16] Alfaro FJ, Lioutas V-A, Pimentel DA, Chung C-C, Bedoya F, Yoo W-K, et al. Cognitive decline in metabolic syndrome is linked to microstructural white matter abnormalities. Journal of Neurology. 2016;**263**:2505-2514

[17] Tsai C-K, Kao T-W, Lee J-T, Wu C-J, Hueng D-Y, Liang C-S, et al. Increased

risk of cognitive impairment in patients with components of metabolic syndrome. Medicine (Baltimore). 2016;**95**(36):e4791

[18] Yates KF, Sweat V, Yau PL, Turchiano MM, Convit A. Impact of metabolic syndrome on cognition and brain: A selected review of the literature. Arteriosclerosis, Thrombosis, and Vascular Biology. 2012;**32**(9):2060-2067

[19] Bayliss WM. On the local reactions of the arterial wall to changes of internal pressure. The Journal of Physiology. 1902;**28**(3):220-231

[20] Schubert R, Kalentchuk VU, Krien U. Rho kinase inhibition partly weakens myogenic reactivity in rat small arteries by changing calcium sensitivity. American Journal of Physiology. Heart and Circulatory Physiology. 2002;**283**:H2288-H2295

[21] Davis MJ, Sikes PJ. Myogenic responses of isolated arterioles: Test for a rate-sensitive mechanism. American Journal of Physiology. Heart and Circulatory Physiology. 1990;**28** (6 Pt 2):H1890-H1900

[22] Folkow B. Transmural pressure and vascular tone—Some aspects of an old controversy. Archives Internationales de Pharmacodynamie et de Thérapie. 1962;**139**:455-469

[23] Mellander S, Johansson B. Control of resistance, exchange, and capacitance functions in the peripheral circulation. Pharmacological Reviews. 1968;**20**(3):117-196

[24] Davis MJ, Hill MA. Signaling mechanisms underlying the vascular myogenic response. Physiological Reviews. 1999;**79**(2):387-423

[25] Roman R, Dokkum R. Commentary on the special issue on the impact of myogenic tone in health and disease. Current Vascular Pharmacology. 2014;**12**(6):779-780

[26] Von der Weid PY. Lymphatic myogenic constriction—How lymphatic vessels pump lymph uphill. The Journal of Physiology. 2013;**591**(2):391-392

[27] Osol G, Brekke JF, Mcelroy-Yaggy K, Gokina NI. Myogenic tone, reactivity, and forced dilatation: A three-phase model of in vitro arterial myogenic behavior. American Journal of Physiology. Heart and Circulatory Physiology. 2002;**283**:H2260-H2267

[28] Harder DR. A cellular mechanism for myogenic regulation of cat cerebral arteries. Annals of Biomedical Engineering. 1985;**13**:335-339

[29] Harder DR, Lange AR, Gebremedhin D, Birks EK, Roman RJ. Cytochrome P450 metabolites of arachidonic acid as intracellular signaling molecules in vascular tissue. Journal of Vascular Research. 1997;**34**(3):237-243

[30] Johnson PC. The myogenic response in the microcirculation and its interaction with other control systems. Journal of Hypertension. Supplement. 1989;7(4):S33-S40

[31] Zou H, Ratz PH, Hill MA. Role of myosin phosphorylation and $[Ca^{2+}]i$ in myogenic reactivity and arteriolar tone. American Journal of Physiology. Heart and Circulatory Physiology. 1995;**269** (5 pt 2):H1590-H1596

[32] Knot HJ, Nelson MT. Regulation of arterial diameter and wall $[Ca^{2+}]$ in cerebral arteries of rat by membrane potential and intravascular pressure. The Journal of Physiology. 1998;**508**(1):199-209

[33] Moosmang S, Schulla V, Welling A, Feil R, Feil S, Rg JÈ, et al. Dominant role of smooth muscle L-type calcium channel $Ca_{v1.2}$ for blood pressure regulation. European Molecular Biology Organization. 2003;**22**(22):6027-6034

[34] Hill MA, Zou H, Potocnik SJ, Meininger GA, Davis MJ. Invited review: Arteriolar smooth muscle mechanotransduction: Ca^{2+} signaling pathways underlying myogenic reactivity. Journal of Applied Physiology. 2001;**91**(16):973-983

[35] Schubert R, Lidington D, Bolz SS. The emerging role of Ca^{2+} sensitivity regulation in promoting myogenic vasoconstriction. Cardiovascular Research. 2008;**77**(1):8-18

[36] Butcher JT, Goodwill AG, Frisbee JC. The ex vivo isolated skeletal microvessel preparation for investigation of vascular reactivity. Journal of Visualized Experiments. 2012;**62**:e3674

[37] Brayden JE, Nelson MT. Regulation of arterial tone by activation of calcium-dependent potassium channels. Science. 1992;**256**(5056):532-535

[38] Knot H, Standen N, Nelson M. Ryanodine receptors regulate arterial diameter and wall [Ca^{2+}] in cerebral arteries of rat via Ca^{2+}-dependent K^+ channels. The Journal of Physiology. 1998;**508**(1):211-221

[39] Brenner R, Pere ÂZGJ, Bonev AD, Eckman DM, Kosek JC, Wiler SW, et al. Vasoregulation by the B1 subunit of the calcium-activated potassium channel. Nature. 2000;**407**:870-876

[40] D'Angelo G, Mogford JE, Davis GE, Davis MJ, Meininger GA. Integrin-mediated reduction in vascular smooth muscle [Ca^{2+}]i induced by RGD-containing peptide. American Journal of Physiology. Heart and Circulatory Physiology. 1997;**272**(4 Pt 2): H2065-H2070

[41] Martinez-Lemus LA, Crow T, Davis MJ, Meininger GA. $\alpha_v\beta_3$- and $\alpha_5\beta_1$-integrin blockade inhibits myogenic constriction of skeletal muscle resistance arterioles. American Journal of Physiology: Heart and Circulatory Physiology. 2005;**289**(1):H322-H329

[42] Lucchesi PA, Sabri A, Belmadani S, Matrougui K. Involvement of metalloproteinases 2/9 in epidermal growth factor receptor transactivation in pressure-induced myogenic tone in mouse mesenteric resistance arteries. Circulation. 2004;**110**:3587-3593

[43] Pires PW, Dams Ramos CM, Matin N, Dorrance AM. The effects of hypertension on the cerebral circulation. American Journal of Physiology. Heart and Circulatory Physiology. 2013;**304**:H1598-H1614

[44] Paterno R, Heistad DD, Faraci FM. Potassium channels modulate cerebral autoregulation during acute hypertension. American Journal of Physiology: Heart and Circulatory Physiology. 2000;**278**:H2003-H2007

[45] Vanbavel E, Van Der Meulen ET, Spaan JAE. Role of rho-associated protein kinase in tone and calcium sensitivity of cannulated rat mesenteric small arteries. Experimental Physiology. 2001;**86**(5):585-592

[46] Lagaud G, Gaudreault N, Moore EDW, Van Breemen C, Laher I. Pressure-dependent myogenic constriction of cerebral arteries occurs independently of voltage-dependent activation. American Journal of Physiology. Heart and Circulatory Physiology. 2002;**283**(6):H2187-H2195

[47] Bolz S-S, Vogel L, Sollinger D, Derwand R, Boer C, Pitson SM, et al. Sphingosine kinase modulates microvascular tone and myogenic responses through activation of RhoA/rho kinase. Circulation. 2003;**108**:342-347

[48] Gokina NI, Park KM, Mcelroy-Yaggy K, Osol G. Biomechanics and mechanotransduction in cells and tissues effects of rho kinase inhibition

on cerebral artery myogenic tone and reactivity. Journal of Applied Physiology. 2005;**98**:1940-1948

[49] Somlyo AP, Somlyo AV. Ca^{2+} sensitivity of smooth muscle and nonmuscle myosin II: Modulated by G proteins, kinases, and myosin phosphatase. Physiological Reviews. 2003;**83**:1325-1358

[50] Osol G, Laher I, Cipolla M. Protein kinase C modulates basal myogenic tone in resistance arteries from the cerebral circulation. Circulation Research. 1991;**68**(2):359-367

[51] Yao L, Brayden JE. Rho kinase activity governs arteriolar myogenic depolarization. Journal of Cerebral Blood Flow and Metabolism. 2017;**37**(1):140-152

[52] Luykenaar KD, Abd El-Rahman R, Walsh MP, Welsh DG. Rho-kinase-mediated suppression of KDR current in cerebral arteries requires an intact actin cytoskeleton. American Journal of Physiology. Heart and Circulatory Physiology. 2009;**296**:H917-H926

[53] Nowicki P, Flavahan S, Hassanain H, Mitra S, Holland S, Goldschmidt-Clemont P, et al. Redox signaling of the arteriolar myogenic response. Circulation Research. 2001;**89**:114-116

[54] Veerareddy S, Cooke C-LM, Baker PN, Davidge ST. Gender differences in myogenic tone in superoxide dismutase knockout mouse: Animal model of oxidative stress. American Journal of Physiology. Heart and Circulatory Physiology. 2004;**287**:H40-H45

[55] Cipolla MJ, Osol G. Vascular smooth muscle actin cytoskeleton in cerebral artery forced dilatation. Stroke. 1998;**29**:1223-1228

[56] Morgan KG, Gangopadhyay SS. Invited review: Cross-bridge regulation by thin filament-associated proteins. Journal of Applied Physiology. 2001;**91**:953-962

[57] Jaggar JH, Porter VA, Lederer WJ, Nelson MT. Calcium sparks in smooth muscle. American Journal of Physiology. Cell Physiology. 2000;**278**(2):C235-C256

[58] Koller A, Toth P. Contribution of flow-dependent vasomotor mechanisms to the autoregulation of cerebral blood flow. Journal of Vascular Research. 2012;**49**(5):375-389

[59] Fujii K, Heistad DD, Faraci FM. Flow-mediated dilatation of the basilar artery in vivo. Circulation Research. 1991;**69**(3):697-705

[60] Thorin-Trescases N, Bevan JA. High levels of myogenic tone antagonize the dilator response to flow of small rabbit cerebral arteries. Stroke. 1998;**29**:1194-1201

[61] Moncada S, Palmer RMJ, Higgs EA. Nitric oxide: Physiology, pathophysiology, and pharmacology. Pharmacological Reviews. 1992;**43**(2):109-142

[62] Paravicini TM, Miller AA, Drummond GR, Sobey CG. Flow-induced cerebral vasodilatation in vivo involves activation of phosphatidylinositol-3 kinase, NADPH-oxidase, and nitric oxide synthase. Journal of Cerebral Blood Flow and Metabolism. 2006;**26**(6):836-845

[63] Sobey CG, Faraci FM. Effects of a novel inhibitor of guanylyl cyclase on dilator responses of mouse cerebral arterioles. Stroke. 1997;**28**(4):837-843

[64] Katsuki S, Arnold W, Mittal C, Murad F. Stimulation of guanylate cyclase by sodium nitroprusside, nitroglycerin and nitric oxide in various tissue preparations and comparison to the effects of sodium azide and hydroxylamine. Journal of Cyclic Nucleotide Research. 1977;**3**(1):23-35

[65] Robertson BE, Schubert R, Hescheler J, Nelson MT. cGMP-dependent protein kinase activates Ca-activated K channels in cerebral artery smooth muscle cells. American Journal of Physiology. Cell Physiology. 1993;**265**(1 pt 1):C299-C303

[66] Porasuphatana S, Tsai P, Rosen GM. The generation of free radicals by nitric oxide synthase. Comparative Biochemistry and Physiology, Part C: Toxicology & Pharmacology. 2003;**134**(3):281-289

[67] Stroes EY, Hijmering M, Zandvoort M, Stroes EY, Hijmering M, Van Zandvoort M, et al. Origin of superoxide production by endothelial nitric oxide synthase. FEBS Letters. 1998;**438**(3):161-164

[68] Drouin A, Thorin E. Flow-induced dilation is mediated by Akt-dependent activation of endothelial nitric oxide synthase-derived hydrogen peroxide in mouse cerebral arteries. Stroke. 2009;**40**:1827-1833

[69] Madden JA, Christman NJT. Integrin signaling, free radicals, and tyrosine kinase mediate flow constriction in isolated cerebral arteries. American Journal of Physiology. 1999;**277**(6):H2264-H2271

[70] Bryan RM, Marrelli SP, Steenberg ML, Schildmeyer LA, Johnson TD. Effects of luminal shear stress on cerebral arteries and arterioles. American Journal of Physiology. Heart and Circulatory Physiology. 2001;**280**:H2011-H2022

[71] Toth P, Rozsa B, Springo Z, Doczi T, Koller A. Isolated human and rat cerebral arteries constrict to increases in flow: Role of 20-HETE and TP receptors. Journal of Cerebral Blood Flow and Metabolism. 2011;**31**:2096-2105

[72] Garcia-Roldan J-L, Bevan JA. Augmentation of endothelium-independent flow constriction in pial arteries at high intravascular pressures. Hypertension. 1991;**17**:870-874

[73] Hoogerwerf N, Zijlstra EJ, van der Linden PJW, Westerhof N, Sipkema P. Endothelium function is protected by albumin and flow-induced constriction is independent of endothelium and tone in isolated rabbit femoral artery. Journal of Vascular Research. 1992;**29**(5):367-375

[74] Hoiland RL, Fisher JA, Ainslie PN. Regulation of the cerebral circulation by arterial carbon dioxide. Comprehensive Physiology. 2019;**9**:1101-1154

[75] Drouin A, Bolduc V, Thorin-Trescases N, Bélanger É, Fernandes P, Baraghis E, et al. Catechin treatment improves cerebrovascular flow-mediated dilation and learning abilities in atherosclerotic mice. American Journal of Physiology. Heart and Circulatory Physiology. 2011;**300**:H1032-H1043

[76] Raignault A, Bolduc V, Lesage F, Thorin E. Pulse pressure-dependent cerebrovascular eNOS regulation in mice. Journal of Cerebral Blood Flow and Metabolism. 2017;**37**(2):413-424

[77] Fujii K, Heistad DD, Faraci FM. Effect of diabetes mellitus on flow-mediated and endothelium-dependent dilatation of the rat basilar artery. Stroke. 1992;**23**(10):1494-1498

[78] Arnold WP, Mittal CK, Katsuki S, Murad F. Nitric oxide activates guanylate cyclase and increases guanosine 3′:5′-cyclic monophosphate levels in various tissue preparations. Proceedings of the National Academy of Sciences of the United States of America. 1977;**74**(8):3203-3207

[79] Faraci FM, Heistad DD. Regulation of large cerebral arteries and cerebral microvascular pressure. Circulation Research. 1990;**66**:8-17

[80] Faraci FM, Mayhan WG, Heistad DD. Segmental vascular responses to acute hypertension in cerebrum and brain stem. American Journal of Physiology. Heart and Circulatory Physiology. 1987;**252** (4 Pt 2):H738-H742

[81] Ngai AC, Winn HR. Modulation of cerebral arteriolar diameter by intraluminal flow and pressure. Circulation Research. 1995;77(4):832-840

[82] Shimoda LA, Norins NA, Jeutter DC, Madden JA. Flow-induced responses in piglet isolated cerebral arteries. Pediatric Research. 1996;**39** (4 Pt 1):574-583

[83] Burnstock G. Neurogenic control of cerebral circulation. Cephalalgia. 1985;**2**:25-33

[84] Jones CJ, Kuo L, Davis MJ, Chilian WM. Myogenic and flow-dependent control mechanisms in the coronary microcirculation. Basic Research in Cardiology. 1993;**88**(1):2-10

[85] Reich T, Rusinek H. Cerebral cortical and white matter reactivity to carbon dioxide. Stroke. 1989;**20**:453-457

[86] de Wit C, Schäfer C, von Bismarck P, Bolz S-S, Pohl U. Elevation of plasma viscosity induces sustained NO-mediated dilation in the hamster cremaster microcirculation in vivo. Pflügers Archiv. European Journal of Physiology. 1997;**434**:354-361

[87] Garcia-Roldan JL, Bevan JA. Flow-induced constriction and dilation of cerebral resistance arteries. Circulation Research. 1990;**66**(5):1445-1448

[88] Peterson EC, Wang Z, Britz G. Regulation of cerebral blood flow. International Journal of Vascular Medicine. 2011;**2011**:8

[89] Berne RM, Richard Winn H, Rubio R. The local regulation of cerebral blood flow. Progress in Cardiovascular Diseases. 1981;**24**(3):243-260

[90] Winn RH, Rubio R, Berne RM. Brain adenosine production in the rat during 60 seconds of ischemia. Circulation Research. 1979;**45**(4):486-492

[91] Bari F, Louis TM, Busija DW. Effects of ischemia on cerebral arteriolar dilation to arterial hypoxia in piglets. Stroke. 1998;**29**:222-228

[92] Ramkumar V, Hallam DM, Nie Z. Adenosine, oxidative stress and cytoprotection. Japanese Journal of Pharmacology. 2001;**86**:265-274

[93] Tabrizchi R, Bedi S. Pharmacology of adenosine receptors in the vasculature. Pharmacology & Therapeutics. 2001;**91**(2):133-147

[94] Ohta M, Toyama K, Gutterman DD, Campbell WB, Lemaître V, Teraoka R, et al. Ecto-5′-nucleotidase, CD73, is an endothelium-derived hyperpolarizing factor synthase. Arteriosclerosis, Thrombosis, and Vascular Biology. 2013;**33**(3):629-636

[95] Wechsler RL, Kleiss LM, Kety SS. The effects of intravenously administered aminophylline on cerebral circulation and metabolism in man. The Journal of Clinical Investigation. 1950;**29**(1):28-30

[96] Gottstein U, Paulson B. The effect of intracarotid aminophylline infusion on the cerebral circulation. Stroke. 1972;**3**:560-565

[97] Magnussen I, Høedt-Rasmussen K. The effect of intraarterial administered aminophylline on cerebral hemodynamics in man. Acta Neurologica Scandinavica. 1977;**55**(2):131-136

[98] Fredholm BB, Ijzerman AP, Jacobson KA, Klotz K-N, Linden J. International Union of Pharmacology.

XXV. Nomenclature and classification of adenosine receptors. Pharmacological Reviews. 2001;**53**(4):527-552

[99] Kalaria RN, Harik SI. Adenosine receptors and the nucleoside transporter in human brain vasculature. Journal of Cerebral Blood Flow and Metabolism. 1988;**8**:32-39

[100] Kalaria RN, Harik SI. Adenosine receptors of cerebral microvessels and choroid plexus. Journal of Cerebral Blood Flow and Metabolism. 1986;**6**:463-470

[101] Phillis JW. Adenosine in the control of the cerebral circulation. Cerebrovascular and Brain Metabolism Reviews. 1989;**1**(1):26-54

[102] Fredholm BB, Abbracchio MP, Burnstock G, Dubyak GR, Harden TK, Jacobson KA, et al. Towards a revised nomenclature for P1 and P2 receptors. Trends in Pharmacological Sciences. 1997;**18**(3):79-82

[103] Brunton L, Chabner B, Knollman B. Goodman and Gilman's: The Pharmacological Basis of Therapeutics. New York City, NY, USA; 2011

[104] Ngai AC, Winn HR. Effects of adenosine and its analogues on isolated intracerebral arterioles extraluminal and intraluminal application. Circulation Research. 1993;**73**:448-457

[105] Anand-Srivastava MB, Franks DJ, Cantin M, Genest J. Presence of "Ra" and "P"-site receptors for adenosine coupled to adenylate cyclase in cultured vascular smooth muscle cells. Biochemical and Biophysical Research Communications. 1982;**108**(1):213-219

[106] de Jong J, de Jonge R, Keijzer E, Bradamante S. The role of adenosine in preconditioning. Pharmacology & Therapeutics. 2000;**87**(2-3):141-149

[107] Kontos HA, Wei EP, Jarrell Raper A, Patterson JL. Local mechanism of CO_2 action on cat pial arterioles. Stroke. 1977;**8**(2):226-229

[108] Li Y-O, Fredholm BB. Adenosine analogues stimulate cyclic AMP formation in rabbit cerebral microvessels via adenosine A2-receptors. Acta Physiologica Scandinavica. 1985;**124**:253-259

[109] Bryan PT, Marshall JM. Adenosine receptor subtypes and vasodilatation in rat skeletal muscle during systemic hypoxia: A role for A1 receptors. The Journal of Physiology. 1999;**514**(1):151-162

[110] Danialou G, Vicaut E, Sambe A, Aubier M, Boczkowski J. Predominant role of A1 adenosine receptors in mediating adenosine induced vasodilatation of rat diaphragmatic arterioles: Involvement of nitric oxide and the ATP-dependent K^+ channels. British Journal of Pharmacology. 1997;**121**:1355-1363

[111] Kuo L, Chancellor JD. Adenosine potentiates flow-induced dilation of coronary arterioles by activating K_{ATP} channels in endothelium. American Journal of Physiology. 1995;**269** (2 pt 2):H541-H549

[112] Li Q, Puro DG. Adenosine activates ATP-sensitive K^+ currents in pericytes of rat retinal microvessels: Role of A1 and A2a receptors. Brain Research. 2001;**907**(1-2):93-99

[113] Carter RW, Kanagy NL. Tyrosine kinases regulate intracellular calcium during 2-adrenergic contraction in rat aorta. American Journal of Physiology. Heart and Circulatory Physiology. 2002;**283**:H1673-H1680

[114] Klinger M, Freissmuth M, Nanoff C. Adenosine receptors: G protein-mediated signalling and the role of accessory proteins. Cellular Signalling. 2002;**14**(2):99-108

[115] Linden J. Molecular approach to adenosine receptors: Receptor-mediated mechanisms of tissue protection. Annual Review of Pharmacology and Toxicology. 2001;**41**:775-787

[116] Gordon GRJ, Choi HB, Ellis-Davies GCR, Macvicar BA. Brain metabolic state dictates the polarity of astrocyte control over the cerebrovasculature. Nature. 2008;**456**(7223):745-749

[117] Morii S, Ngai AC, Ko KR, Winn HR. Role of adenosine in regulation of cerebral blood flow: Effects of theophylline during normoxia and hypoxia. American Journal of Physiology. Heart and Circulatory Physiology. 1987;**253**(1 pt 2):H165-H175

[118] Willie CK, Tzeng Y-C, Fisher JA, Ainslie PN. Integrative regulation of human brain blood flow. The Journal of Physiology. 2014;**592**(5):841-859

[119] Ainslie PN, Ashmead JC, Ide K, Morgan BJ, Poulin MJ. Differential responses to CO_2 and sympathetic stimulation in the cerebral and femoral circulations in humans. The Journal of Physiology. 2005;**566**(2):613-624

[120] Willie CK, Macleod DB, Shaw AD, Smith KJ, Tzeng YC, Eves ND, et al. Regional brain blood flow in man during acute changes in arterial blood gases. The Journal of Physiology. 2012;**590**(12):3261-3275

[121] Giller CA, Bowman G, Dyer H, Mootz L, Krippner W. Cerebral arterial diameters during changes in blood pressure and carbon dioxide during craniotomy. Neurosurgery. 1993;**32**(5):737-742

[122] Wilson MH, Edsell ME, Davagnanam I, Hirani SP, Martin DS, Levett DZ, et al. Cerebral artery dilatation maintains cerebral oxygenation at extreme altitude and in acute hypoxia—An ultrasound and MRI study. Journal of Cerebral Blood Flow and Metabolism. 2011;**31**:2019-2029

[123] Willie CK, Smith KJ, Day TA, Ray LA, Lewis NCS, Bakker A, et al. Regional cerebral blood flow in humans at high altitude: Gradual ascent and 2 wk at 5,050 m. Journal of Applied Physiology. 2014;**116**:905-910

[124] Wolff HG. Cerebral circulation. Archives of Neurology and Psychiatry. 1930;**23**(6):1097

[125] Binks AP, Cunningham VJ, Adams L, Banzett RB. Gray matter blood flow change is unevenly distributed during moderate isocapnic hypoxia in humans. Journal of Applied Physiology. 2008;**104**:212-217

[126] Mandell DM, Han JS, Poublanc J, Crawley AP, Kassner A, Fisher JA, et al. Selective reduction of blood flow to white matter during hypercapnia corresponds with leukoaraiosis. Stroke. 2008;**39**:1993-1998

[127] Nöth U, Kotajima F, Delchmann R, Turner R, Corfield DR. Mapping of the cerebral vascular response to hypoxia and hypercapnia using quantitative perfusion MRI at 3T. NMR in Biomedicine. 2008;**21**(3):464-472

[128] Piechnik SK, Chiarelli PA, Jezzard P. Modelling vascular reactivity to investigate the basis of the relationship between cerebral blood volume and flow under CO_2 manipulation. NeuroImage. 2008;**39**:107-118

[129] Betz E, Enzenrob HG, Vlahov V. Interaction of H^+ and Ca^{++} in the regulation of local pial vascular resistance. Pflfigers Archive. 1973;**343**:79-88

[130] Liu X, Li C, Falck JR, Harder DR, Koehler RC. Relative contribution of cyclooxygenases, epoxyeicosatrienoic acids, and pH to the cerebral blood

flow response to vibrissal stimulation. American Journal of Physiology. Heart and Circulatory Physiology. 2012;**302**:H1075-H1085

[131] Wahl M, Deetjen P, Thurau K, Ingvar DH, Lassen NA. Micropuncture evaluation of the importance of perivascular pH for the arteriolar diameter on the brain surface. Pflügers Archiv. 1970;**316**:152-163

[132] Kontos HA, Raper AJ, Patterson JL. Analysis of vasoactivity of local pH, PCO_2 and bicarbonate on pial vessels. Stroke. 1977;**8**(3):358-360

[133] Harper AM, Bell RA. The effect of metabolic acidosis and alkalosis on the blood flow through the cerebral cortex. Journal of Neurology, Neurosurgery, and Psychiatry. 1963;**26**(4):341-344

[134] Koehler RC, Traystman RJ. Bicarbonate ion modulation of cerebral blood flow during hypoxia and hypercapnia. American Journal of Physiology. 1982;**243**(1):H33-H40

[135] Boedtkjer E, Hansen KB, Boedtkjer DMB, Aalkjaer C, Boron WF. Extracellular HCO_3^- is sensed by mouse cerebral arteries: Regulation of tone by receptor protein tyrosine phosphatase y. Journal of Cerebral Blood Flow and Metabolism. 2015;**36**(5):965-980

[136] Chen Y, Cann MJ, Litvin TN, Iourgenko V, Sinclair ML, Levin LR, et al. Soluble adenylyl cyclase as an evolutionarily conserved bicarbonate sensor. Science. 2000;**289**(5479):625-628

[137] Busse R, Förstermann U, Matsuda H, Pohp U. The role of prostaglandins in the endothelium-mediated vasodilatory response to hypoxia. Pflügers Archiv. 1984;**401**:77-83

[138] Lynch FM, Austin C, Heagerty AM, Izzard AS. Adenosine and hypoxic dilation of rat coronary small arteries: Roles of the ATP-sensitive potassium channel, endothelium and nitric oxide. American Journal of Physiology. Heart and Circulatory Physiology. 2006;**290**:H1145-H1150

[139] Liu Q, Flavahan NA. Hypoxic dilatation of porcine small coronary arteries: Role of endothelium and K_{ATP} channels. British Journal of Pharmacology. 1997;**120**:728-734

[140] Messina EJ, Sun D, Koller A, Wolin MS, Kaley G. Role of endothelium-derived prostaglandins in hypoxia-elicited arteriolar dilation in rat skeletal muscle. Circulation Research. 1992;**71**:790-796

[141] Frisbee JC, Maier KG, Falck JR, Roman RJ, Lombard JH. Integration of hypoxic dilation signaling pathways for skeletal muscle resistance arteries. American Journal of Physiology. Regulatory, Integrative and Comparative Physiology. 2002;**283**:R309-R319

[142] Myers PR, Banitt PF, Guerra R, Harrison DG. Role of the endothelium in modulation of the acetylcholine vasoconstrictor response in porcine coronary microvessels. Cardiovascular Research. 1991;**25**:129-137

[143] Myers PR, Muller JM, Tanner MA. Effects of oxygen tension on endothelium dependent responses in canine coronary microvessels. Cardiovascular Research. 1991;**25**:885-894

[144] Busse R, Pohl U, Kellner C, Klemm U. Endothelial cells are involved in the vasodilatory response to hypoxia. Pflügers Archiv. 1983;**397**:78-80

[145] Halvorson BD, Whitehead SN, Mcguire JJ, Wiseman RW, Frisbee JC. Endothelium-dependent impairments to cerebral vascular reactivity with type 2 diabetes mellitus in the Goto-Kakizaki

rat. American Journal of Physiology. Regulatory, Integrative and Comparative Physiology. 2019;**317**:R149-R159

[146] Van Mil AHM, Spilt A, Van Buchem MA, Bollen ELEM, Teppema L, Westendorp RGJ, et al. Nitric oxide mediates hypoxia-induced cerebral vasodilation in humans. Journal of Applied Physiology. 2002;**92**:962-966

[147] MacVicar BA, Newman EA. Astrocyte regulation of blood flow in the brain. Cold Spring Harbor Perspectives in Biology. 2015;**7**:1-15

[148] Faraci FM, Sobey CG. Role of potassium channels in regulation of cerebral vascular tone. Journal of Cerebral Blood Flow and Metabolism. 1998;**18**:1047-1063

[149] Rosenblum WI. ATP-sensitive potassium channels in the cerebral circulation. Stroke. 2003;**34**:1547-1552

[150] Filosa JA, Bonev AD, Straub SV, Meredith AL, Wilkerson K, Aldrich RW, et al. Local potassium signaling couples neuronal activity to vasodilation in the brain. Nature Neuroscience. 2006;**9**(11):1397-1403

[151] Villringer A, Dirnagl U. Coupling of brain activity and cerebral blood flow: Basis of functional neuroimaging. Cerebrovascular and Brain Metabolism Reviews. 1995;**7**(3):240-276

[152] Lecrux C, Hamel E. The neurovascular unit in brain function and disease. Acta Physiologica. 2011;**203**(1):47-59

[153] Hyder F. Dynamic imaging of brain function. Methods in Molecular Biology. 2009;**489**:3-21

[154] Schlageter KE, Molnar P, Lapin GD, Groothuis DR. Microvessel organization and structure in experimental brain tumors: Microvessel populations with distinctive structural and functional properties. Microvascular Research. 1999;**58**(3):312-328

[155] Simard M, Arcuino G, Takano T, Liu QS, Nedergaard M. Signaling at the gliovascular interface. The Journal of Neuroscience. 2003;**23**(27):9254-9262

[156] Figueroa XF, Duling BR. Gap junctions in the control of vascular function. Antioxidants & Redox Signaling. 2009;**11**(2):251-266

[157] Del Zoppo GJ, Milner R. Integrin-matrix interactions in the cerebral microvasculature. Arteriosclerosis, Thrombosis, and Vascular Biology. 2006;**26**(9):1966-1975

[158] Brayden JE, Earley S, Nelson MT, Reading S. Transient receptor potential (TRP) channels, vascular tone and autoregulation of cerebral blood flow. Clinical and Experimental Pharmacology & Physiology. 2008;**35**:1116-1120

[159] Attwell D, Mishra A, Hall CN, O'Farrell FM, Dalkara T. What is a pericyte? Journal of Cerebral Blood Flow and Metabolism. 2016;**36**(2):451-455

[160] Lecrux C, Hamel E. Neuronal networks and mediators of cortical neurovascular coupling responses in normal and altered brain states. Philosophical Transactions of the Royal Society B: Biological Sciences. 2016;**371**(1705):1-13

[161] Iadecola C. The neurovascular unit coming of age: A journey through neurovascular coupling in health and disease. Neuron. 2017;**96**(1):17-42

[162] Sokoya EM, Burns AR, Setiawan CT, Coleman HA, Parkington HC, Tare M. Evidence for the involvement of myoendothelial gap junctions in EDHF-mediated relaxation in the rat middle cerebral

artery. American Journal of Physiology. Heart and Circulatory Physiology. 2006;**291**:H385-H393

[163] Hannah RM, Dunn KM, Bonev AD, Nelson MT. Endothelial SK_{Ca} and IK_{Ca} channels regulate brain parenchymal arteriolar diameter and cortical cerebral blood flow. Journal of Cerebral Blood Flow and Metabolism. 2011;**31**(5):1175-1186

[164] Nguyen T, Richard Winn H, Janigro D. ATP-sensitive potassium channels may participate in the coupling of neuronal activity and cerebrovascular tone. American Journal of Physiology. Heart and Circulatory Physiology. 2000;**278**:H878-H885

[165] McCarron JG, Halpern W. Potassium dilates rat cerebral arteries by two independent mechanisms. American Journal of Physiology. 1990;**259**(3 Pt 2):H902-H908

[166] Knot HJ, Zimmermann PA, Nelson MT. Extracellular K^+-induced hyperpolarizations and dilatations of rat coronary and cerebral arteries involve inward rectifier K^+ channels. The Journal of Physiology. 1996;**492**(2):419-430

[167] Zaritsky JJ, Eckman DM, Wellman GC, Nelson MT, Schwarz TL. Targeted disruption of Kir2.1 and Kir2.2 genes reveals the essential role of the inwardly rectifying K current in K-mediated vasodilation. Circulation Research. 2000;**87**:160-166

[168] Longden TA, Dabertrand F, Koide M, Gonzales AL, Tykocki NR, Brayden JE, et al. Capillary K^+-sensing initiates retrograde hyperpolarization to increase local cerebral blood flow. Nature Neuroscience. 2017;**20**(5):717-726

[169] Somjen GG. Mechanisms of spreading depression and hypoxic spreading depression-like depolarization. Physiological Reviews. 2001;**81**(3):1065-1096

[170] Vyskočil F, Kríž N, Bureš J. Potassium-selective microelectrodes used for measuring the extracellular brain potassium during spreading depression and anoxic depolarization in rats. Brain Research. 1972;**39**:255-259

[171] Haddy FJ, Vanhoutte PM, Feletou M. Role of potassium in regulating blood flow and blood pressure. American Journal of Physiology. Regulatory, Integrative and Comparative Physiology. 2006;**290**:R546-R552

[172] Edwards FR, Hirst GDS, Silverberg GD. Inward rectification in rat cerebral arterioles; involvement of potassium ions in autoregulation. The Journal of Physiology. 1988;**404**:455-466

[173] Quayle JM, Mccarron JG, Brayden JE, Nelson MT. Inward rectifier K^+ currents in smooth muscle cells from rat resistance-sized cerebral arteries. American Journal of Physiology. 1993;**265**(5 pt 1):C1363-C1370

[174] Robertson BE, Bonev AD, Nelson MT. Inward rectifier K^+ currents in smooth muscle cells from rat coronary arteries: Block by Mg^{2+}, Ca^{2+}, and Ba^{2+}. American Journal of Physiology. 1996;**271**(2 pt 2): H696-H705

[175] Bünger R, Francis JH, Querengässer A, Gerlach E. Studies on potassium induced coronary dilation in the isolated guinea pig heart. Pflügers Archiv. 1976;**363**:27-31

[176] Haddy F. Potassium effects on contraction in arterial smooth muscle mediated by Na^+, K^+-ATPase. Federation Proceedings. 1983;**42**(2):239-245

[177] Hermsmeyer K. Sodium pump hyperpolarization-relaxation in rat

caudal artery. Federation Proceedings. 1987;**42**(2):246-252

[178] Burns WR, Cohen KD, Jackson WF. K^+-induced dilation of hamster cremasteric arterioles involves both the Na^+/K^+-ATPase and inward-rectifier K^+ channels. Microcirculation. 2004;**11**:279-293

[179] Frisbee JC, Maier KG, Stepp DW. Oxidant stress-induced increase in myogenic activation of skeletal muscle resistance arteries in obese Zucker rats. American Journal of Physiology. Heart and Circulatory Physiology. 2002;**283**:H2160-H2168

[180] Hayashi K, Kanda T, Homma K, Tokuyama H, Okubo K, Takamatsu I, et al. Altered renal microvascular response in Zucker obese rats. Metabolism. 2002;**51**(12):1553-1561

[181] Jarajapu YPR, Guberski DL, Grant MB, Knot HJ. Myogenic tone and reactivity of cerebral arteries in type II diabetic BBZDR/Wor rat. European Journal of Pharmacology. 2008;**579**:298-307

[182] Phillips SA, Sylvester FA, Frisbee JC. Oxidant stress and constrictor reactivity impair cerebral artery dilation in obese Zucker rats. American Journal of Physiology. Regulatory, Integrative and Comparative Physiology. 2005;**288**:R522-R530

[183] Brooks SD, DeVallance E, D'Audiffret AC, Frisbee SJ, Tabone LE, Shrader CD, et al. Metabolic syndrome impairs reactivity and wall mechanics of cerebral resistance arteries in obese Zucker rats. American Journal of Physiology. Heart and Circulatory Physiology. 2015;**309**(11):H1846-H1859

[184] Baumbach GL, Faraci FM, Heistad DD. Effects of local reduction in pressure on endothelium-dependent responses of cerebral arterioles. Stroke. 1994;**25**:1456-1462

[185] Barbato JE, Zuckerbraun BS, Overhaus M, Raman KG, Tzeng E. Nitric oxide modulates vascular inflammation and intimal hyperplasia in insulin resistance and the metabolic syndrome. American Journal of Physiology. Heart and Circulatory Physiology. 2005;**289**:H228-H236

[186] Dandona P, Aljada A, Chaudhuri A, Mohanty P, Garg R. Metabolic syndrome: A comprehensive perspective based on interactions between obesity, diabetes, and inflammation. Circulation. 2005;**111**:1448-1454

[187] Kazuyama E, Saito M, Kinoshita Y, Satoh I, Dimitriadis F, Satoh K. Endothelial dysfunction in the early- and late-stage type-2 diabetic Goto-Kakizaki rat aorta. Molecular and Cellular Biochemistry. 2009;**332**:95-102

[188] Cipolla MJ, Liebeskind DS, Chan SL. The importance of comorbidities in ischemic stroke: Impact of hypertension on the cerebral circulation. Journal of Cerebral Blood Flow and Metabolism. 2018;**38**(12):2129-2149

[189] González JM, Beatriz S, Conde MV, Fernández-Alfonso MS, González MC, Arribas SM. Hypertension increases middle cerebral artery resting tone in spontaneously hypertensive rats: Role of tonic vasoactive factor availability. Clinical Science. 2008;**114**:651-659

[190] Bauersachs J, Bouloumié A, Mülsch A, Wiemer G, Fleming I, Busse R. Vasodilator dysfunction in aged spontaneously hypertensive rats: Changes in NO synthase III and soluble guanylyl cyclase expression, and in superoxide anion production. Cardiovascular Research. 1998;**37**:772-779

[191] López-Farré A, Rodriguez-Feo JA, García-Colis E, Gomez J, López-Blaya A, Fortes J, et al. Reduction of

the soluble cyclic GMP vasorelaxing system in the vascular wall of stroke-prone spontaneously hypertensive rats: Effect of the α1-receptor blocker doxazosin. Journal of Hypertension. 2002;**20**(3):463-470

[192] Ruetten H, Zabel U, Linz W, Schmidt HHHW. Downregulation of soluble guanylyl cyclase in young and aging spontaneously hypertensive rats. Circulation Research. 1999;**85**(6):534-541

[193] Humphrey JD. Mechanisms of arterial remodeling in hypertension coupled roles of wall shear and intramural stress. Hypertension. 2008;**52**(2):195-200

[194] Davies PF, Polacek DC, Handen JS, Helmke BP, DePaola N. A spatial approach to transcriptional profiling: Mechanotransduction and the focal origin of atherosclerosis. Trends in Biotechnology. 1999;**17**(9):347-351

[195] Davies PF. Hemodynamic shear stress and the endothelium in cardiovascular pathophysiology. Nature Clinical Practice. Cardiovascular Medicine. 2009;**6**(1):16-26

[196] Heistad DD, Baumbach GL. Cerebral vascular changes during chronic hypertension: Good guys and bad guys. Journal of Hypertension. Supplement. 1992;**10**(7):S71-S75

[197] Iadecola C, Davisson RL. Hypertension and cerebrovascular dysfunction. Cell Metabolism. 2008;**7**(6):476-484

[198] Chantler PD, Frisbee JC. Arterial function in cardio-metabolic diseases: From the microcirculation to the large conduits. Progress in Cardiovascular Diseases. 2015;**57**:489-496

[199] Harris AK, Hutchinson JR, Sachidanandam K, Johnson MH, Dorrance AM, Stepp DW, et al. Type 2 diabetes causes remodeling of cerebrovasculature via differential regulation of matrix metalloproteinases and collagen synthesis role of endothelin-1. Diabetes. 2005;**54**:2638-2644

[200] Payne RA, Wilkinson IB, Webb DJ. Arterial stiffness and hypertension: Emerging concepts. Hypertension. 2010;**55**(1):9-14

Chapter 3

Primary Cilia are Sensory Hubs for Nitric Oxide Signaling

Sidney T. Ley and Wissam A. AbouAlaiwi

Abstract

Primary cilia are sensory organelles present on the surface of most polarized cells. Primary cilia have been demonstrated to play many sensory cell roles, including mechanosensory and chemosensory functions. We demonstrated previously that primary cilia of vascular endothelial cells will bend in response to fluid shear stress, which leads to the biochemical production and release of nitric oxide. This process is impaired in endothelial cells that lack primary cilia function or structure. In this chapter, we will provide an overview of ciliogenesis and the differences between primary cilia and multicilia, as well as an overview of our published work on primary cilia and nitric oxide, and a brief perspective on their implications in health and disease.

Keywords: primary cilia, nitric oxide, signaling, fluid shear stress, mechanosensory transduction

1. Introduction

Cilia are found in nearly every cell in the animal body, where they function as highly specialized sensory organelles. Ciliary malfunction, therefore, tends to result in severe abnormalities, which are often multisystemic. These abnormalities are known as ciliopathies, and as our understanding of cilia form and function continues to grow, so too does the list of known ciliopathies. It is now known that mutations in over 40 genes can alter cilia structure or function, and this list continues to grow; over 1000 polypeptides in the ciliary proteome have yet to be researched [1, 2].

The field of cilia research gained interest after the discovery that cilia play a role in the pathogenesis of Polycystic Kidney Disease (PKD) as fluid mechanosensors within the kidney. In addition to renal dysfunction, the cardiovascular system is also affected by PKD, which has prompted further research into the role that primary cilia play within this system. In kidney tubule epithelia, primary cilia activation leads to a calcium influx, and it has been proposed that this may also occur in vascular endothelial cells. In their study, Nauli et al. showed that vascular cilia play a similar function in sensing fluid shear stress, and there was a corresponding increase in calcium levels correlated with nitric oxide (NO) release. This is thought to contribute to blood pressure control directly. Testing this hypothesis, Nauli et al. showed that cilia mutant cell lines had little to no calcium influx, as well as a lack of NO release while under fluid shear stress [1, 3, 4].

Nitric oxide is a signaling molecule that plays many important functional roles in almost every organ system in the body. Various pathologies are associated with wayward NO production and altered bioavailability levels caused by abnormal

IntechOpen

signaling cascades, which are often the result of abnormal cilia-regulated signaling pathways. There is a documented connection between cilia and NO in the vasculature, as well as an overlap between signaling pathways in other pathologies. It has been postulated that there is a connection between primary cilia and NO outside of the vasculature, but literature on the subject is scarce [1].

This chapter aims to explain cilia type, structure, and function, as well as ciliogenesis, nitric oxide signaling, and finally the interplay between nitric oxide and primary cilia.

2. Cilia type and structure

To understand what makes primary cilia unique, it is important to understand the differences between cilia form and function. Cilia are dynamic sensory organelles present in nearly every cell in every animal, as well as most protozoa. There are two classes of cilia; motile, which possess the dynein motor complexes needed to move, and nonmotile. Motile and nonmotile cilia both contain a 25 μm diameter cytoskeletal scaffold known as the axoneme [5]. The axoneme is comprised of hundreds of proteins and houses nine peripheral microtubule doublets. These doublets are made up of A and B tubules, and they either surround a central pair of microtubules (9 + 2 pattern), or do not (9 + 0 pattern) [5]. Some motile cilia contain a 9 + 2 pattern and exist in clusters on cells called multiciliated cells (MCCs) [6]. There is also a class of motile cilia that have a 9 + 0 structure and exist as solitary monocilia on cell surfaces. The presence or absence of the central pair leads to significant functional differences in the cilia. The 9 + 2 structure commonly moves in a wave-like motion to move fluid, and an example of this are the ependymal cilia. The 9 + 2 patterned cilia also move cerebral spinal fluid, while the 9 + 0 structured most commonly moves in a rotary or corkscrew motion, as seen in flagella, which is useful for propulsion [7, 8]. There is some debate on whether sperm tail flagella should be classified as motile monocilia; regardless, they also possess a similar axonemal structure [5, 6, 9–11]. Nonmotile cilia, known as primary cilia, have a 9 + 0 structure and exist as monocilia on the surface of cells. As primary cilia can be found on vascular endothelial cells, they will be the focus of this review, but a brief overview of multicilia and their motion will also be covered.

2.1 Ciliogenesis

Cilia formation is known as ciliogenesis. Ciliogenesis is correlated with cell division and occurs at the G1/G0 phase of the cell cycle. Reabsorption or disassembly of the cilium starts after cell cycle re-entry. In the first step of ciliogenesis, the centrosome travels to the cell surface, whereupon a basal body is formed by the mother centriole, and it nucleates the ciliary axoneme at the G1/G0 phase of the cell cycle [12]. This first process is regulated by distal appendage proteins, such as centrosomal protein 164 [13]. During the second step, the cilium elongates; this process is regulated by nuclear distribution gene E homolog 1 (Nde1), up until the cilium is matured [14]. The third step is cilia resorption, followed by axonemal shortening during cell cycle reentry. This third process is controlled by the Aurora A-HDAC6, Nek2-Kif24, and Plk1-Kif2A pathways [15]. In the fourth step, the basal body is released from the cilia, which frees the centrioles that act as microtubule organizing centers (MTOC) or spindle poles for mitosis [12].

Immotile cilia formation is impacted by the coordination of the assembly and disassembly equilibrium, the IFT system, and membrane trafficking. When the axoneme nucleates from the basal body, it contains a microtubule bundle contained within the ciliary membrane [16]. Enclosed within are certain signaling molecules and ion channels. Because cilia lack the machinery needed to synthesize ciliary proteins, proteins synthesized by the cell's Golgi apparatus must be transported through a ciliary 'gate' and transition zone near the cilium base [17]. The transition zone, recognizable by a change from triplet to doublet microtubules, is located at the distal end of the basal body (**Figure 1**) [18]. Basal body docking with the plasma membrane can be either permanent, in the case of unicellular organisms, or temporary, in the case of metazoans [5].

Transition fibers, which are present in unicellular organisms, or distal and subdistal appendages, which are present in mammals, are attached to microtubules within the transition zone [19]. Transition fibers function as docking sites for intraflagellar transport (IFT) proteins [20]. IFT transports cargo in a bidirectional manner along the length of cilia, and is mediated by kinesin-2 (anterograde) and cytoplasmic dynein-2 motors (retrograde) attached to multisubunit protein complexes known as IFT particles [21, 22]. Y-linkers exist at the distal end of the transition zone and secure the doublet microtubules to the ciliary membrane in most organisms [19].

2.2 Multicilia

In vertebrates, MCCs are present in a wide variety of different tissue types. In mammals, ependymal MCCs line brain ventricles and the airway epithelium. Multicilia are produced by specialized cells for highly specialized functions. MCCs are typically defined as having more than two cilia on their surface, although this occurrence is not well documented or understood. Recently, MCCs have been observed in unicellular eukaryotes and protists, as well as many metazoans, and even in certain plant sperms [23–25]. MCCs result in the production of motile axonemes, with the only notable exception being mammalian olfactory cilia. These olfactory MCCs lack dynein arms and are considered immotile despite having a 9 + 2

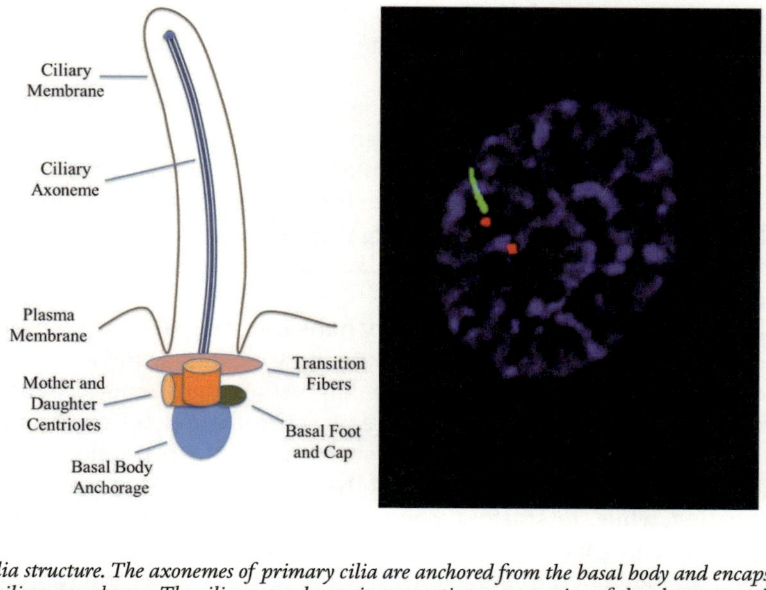

Figure 1.
Primary cilia structure. The axonemes of primary cilia are anchored from the basal body and encapsulated within the ciliary membrane. The ciliary membrane is one continuous extension of the plasma membrane. The basal body is composed of the mother and the daughter centrioles, as well as some transition fibers that anchor the basal body to the cell membrane. The ciliary membrane houses specific membrane and protein receptors, all of which facilitate proper cilia signaling (left panel). Primary cilia that are found on vascular endothelial cells are identifiable by an immunofluorescence technique with antibody against acetylated α-tubulin (green) labeling primary cilia, and pericentrin (red) labeling the centriole or basal body. The nucleus is counterstained with DAPI (blue) to label DNA (right panel). Left panel is adopted from Ref. [1].

structure. This occurrence is indicative of MCCs being a solution to the need for local fluid flow, possibly due to their ability for hydrodynamic coupling [6, 26, 27].

Multicilia carry out their functions by beating, and the basic machinery and organization of cilia beating seems well conserved between eukaryotes, as well as between single motile cilium and multicilia. Some parameters, such as beat frequency, are under cellular control and are varied among cell types. In addition, only motile cilia and sperm flagella contain the dynein machinery needed to power axonemal beating during ATP hydrolysis [5, 28]. The ciliary beat cycle has two phases: the effective stroke, and the recovery stroke. The effective stroke is the initially bending from its upright position, while the recovery stroke sees it return to its original, unbent position [6, 29]. Ciliary motility is controlled by outer and inner axonemal dynein arms, which slide adjacent doublets in respect to one another. The sliding is mediated by protein bridges between doublets, and by the basal anchoring of the axoneme. As a result, cilia bend [6, 29].

The phenomenon metachrony occurs when cilia are organized in such a way that each cilium, in a two-dimensional array, will beat at the same frequency, but in a phase shifted manner. As a result, a traveling wave of ciliary action moves across the array, which propels fluids in a current. Even if each cilium in an array start off in synch, hydrodynamic forces between each cilium will nudge them back towards metachrony, possibly because in a metachronal array, the work each cilium must do is reduced, and more fluid is displaced. Because of this, multiciliation is thought to be a more evolutionarily efficient way to generate fluid flow [6, 30–32].

2.3 Primary cilia

Nearly all human cells house a single nonmotile cilium on their surface, and these primary cilia serve sensory and signaling purposes. The role of primary cilia function and formation on animal health and pathophysiology has only recently been brought to researcher's attentions and is even now not fully understood. A wealth of new information about the primary cilia has been discovered within the last few decades, shedding some light on the function of the previously thought vestigial organelle [33].

Primary cilia are part of various signaling pathways in vertebrates, and usually lack a microtubule central pair (9 + 0 axoneme), as well as the structures associated with the central doublet pair, such as the inner and outer dynein arms required for ciliary movement [34, 35]. Primary cilia can be found in, but are not limited to; endothelia, epithelia, and neurons. Although found in nearly every mammalian cell, notable exceptions include the intercalated cells of the kidney collecting ducts, red blood cells, hepatocytes, and MCCs [34].

Primary cilia function mainly as sensory hubs and are host to many different groups of mechanosensory proteins, chemosensory receptors, and ion channels. These translate extracellular stimuli into an intracellular biochemical signal, which cause different cellular responses. There are two current models that attempt to explain how primary cilia function; the compartment model, and the scaffold model. The compartment model states that the ciliary structure is essential for proper signaling, while the scaffold model states that IFT molecules must play a part in either signaling itself, or the acquisition of outside transduction intermediates [36, 37]. As stated in the ciliogenesis section, primary cilia form through IFT, which transports proteins along the microtubules of the axoneme. The axoneme acts like a scaffold for certain protein complexes, including kinesins and dyneins, which facilitate back-and-forth trafficking of cargo proteins along the length of the cilium. Along with their creation, IFT is also required for maintenance of the primary cilium, and possibly even their core functions [1, 10].

The ciliary membrane is an extension of the cellular membrane where a host of proteins and receptors are housed due to the ciliary transition zone, which provides docking sites for molecular transport into and out of the cilioplasm (**Figure 1**) [38–40]. While there is no confirmed mechanism by which molecules enter and exit the cilia, several mechanisms have been proposed. One such mechanism is the active transport of vesicles from the Golgi apparatus to docking sites in the transition zone [41]. The vesicles are thought to interact with exocyst complexes, where they experience soluble N-ethylmaleimide-sensitive factor (NSF) attachment protein receptor (SNARE) mediated diffusion across the cilioplasm/cytoplasm barrier [37]. The BBSome, which resides in the basal body and is an octameric protein complex, is involved in the movement of transmembrane proteins to the ciliary membrane. BBSomes are known to recognize ciliary targeting sequences and will readily interact with molecules that are upstream of Rab8 activation. BBSomes are not thought to be required for any aspect of ciliogenesis; nevertheless, if BBSomes fail to deliver certain vital proteins to the cilia, the cilia may lose functionality [42–45]. Another proposed mechanism of molecule trafficking is the action of transmembrane proteins, of which some are associated with specific protein sequences that target cilia localization, such as the N-terminal RVxP sequence on polycystin-2 (PC-2) [42, 46].

Primary cilia serve many mechanosensory functions within the body. For example, proper kidney function depends on regulated fluid flow through the nephrons and collecting ducts, which controls the glomerular filtration rate [47]. This flow is sensed by the epithelial primary cilia present in the kidney. Fluid redirection by the primary cilia causes an increase in intracellular calcium. This calcium influx is also mediated by the PC1/2 ion channel complex, and both the mechanosensitive polycystin-1 (PC-1) membrane protein and the PC-2 channel localize to the primary cilia [48–50]. In renal cells, defects in this ion channel complex, along with complete disruption of cilia formation, is known to result in PKD [51].

3. Nitric oxide

Nitric oxide is a signaling molecule involved in a wide array of cellular pathways; mainly, NO contributes to the normal functions of a variety of organ systems [52]. NO is highly reactive, and readily diffuses across cellular membranes. As a result, NO is found in many paracrine signaling pathways. NO is mainly synthesized from l-arginine, oxygen, and NADPH in a redox reaction, catalyzed by nitric oxide synthase (NOS) [53]. NOS has three isoforms, but only endothelial NOS (eNOS) and neuronal NOS (nNOS) are constantly and consistently expressed in cells. Both eNOS and nNOS are calcium dependent, while the other NOS isoform, cytokine inducible NOS (iNOS), is expressed by pro-inflammatory cytokines on an as needed basis [54]. iNOS and nNOS are both soluble enzymes that exist within the cytosol. eNOS, however, is found to localize to the plasma or Golgi body membranes. Because of its unique and wide cellular and subcellular distribution, NOS has many diverse functions throughout the entirety of the body [1, 54, 55].

4. Cilia and nitric oxide interplay

Although primary cilia and NO have various independent roles within the body, especially in the vasculature, their functions often intersect and cooperate with each other. Most research on the interaction between endothelial primary cilia and nitric oxide focuses on vascular homeostasis, but their interactions extend into other areas [1]. However, this chapter will focus on signaling cascades that lead to NO biosynthesis or increased NO bioavailability. The following discussion of the interactions between primary cilia and nitric oxide will focus on vasodilation, wound healing, dopamine signaling, and cellular proliferation.

4.1 Vasodilation

Primary cilia and NO independently effect the vasculature in different ways, but recent studies suggest a direct relationship may exist between the two. Vascular endothelial cells are present in the blood vessel wall and are in continuous contact with blood flow-generated fluid shear stress. Endothelial cells are known mechanotransducers of fluid shear stress, which causes the biosynthesis of NO. This helps regulate vascular tone; NO will diffuse into the surrounding smooth muscle, producing vasorelaxation [56].

Evidence supports primary cilia as the main sensor in this mechanosensitive pathway. As stated previously, PC-1, a mechanosensory protein that malfunctions in polycystic kidney disease, localizes to vascular endothelial primary cilia. In an *in vitro* study performed by Nauli et al., which investigated PC-1's fluid shear mechanosensory properties, it was found that in contrast to wildtype endothelial cells, the PC-1 knockout cells failed to produce an increase in cytosolic calcium and the corresponding NO flux in response to fluid shear stress. The authors, to demonstrate that calcium and NO signals are induced in response to ciliary PC-1 activation, used *Tg737*$^{orpk/orpk}$ endothelial cells that lack ciliary ultrastructure but have functional PC-1. The results showed that neither calcium nor NO signals were present at flow rates up to 50 dyne/cm^2 [3]. These results suggest that PC-1 is responsible for proper cilia mechanosensory function, and that ciliary PC-1 specifically elicits NO production.

Follow-up studies by AbouAlaiwi et al. Showed that PC-2, which, as stated previously, is a calcium permeable cation channel that forms a complex with PC-1, is also important for mechanotransduction. Studies using a PC-2 knockdown line of cells showed a reduction in calcium and NO flux under shear stress when compared to control cells. This was further validated in *ex vivo* studies, where endothelial cells isolated from *pkd2*$^{-/-}$ mice arteries failed to respond to fluid shear stress [4]. These results indicate both PC-1 and PC-2 are needed for cilia mechanosensation, and further suggest that activation of the PC-1/PC-2 complex will start the signaling cascade needed for calcium-dependent NO biosynthesis. In addition, the results show that the increase in intracellular calcium is caused by an increase in intra-ciliary calcium. However, other researchers have proposed that calcium moves bidirectionally between the cilia and the cytosol [57–59]. Regardless, the increase in intracellular calcium triggers the calcium/calmodulin complex, which activates constitutive NOS, such as eNOS, by binding to its target site on the enzyme [60].

The calcium/calmodulin complex can also indirectly activate eNOS through activation of the AKT/PKB pathway, which stimulates AMPK (**Figure 2**) [61]. eNOS activation is mainly calcium dependent, but some studies have shown that a calcium independent pathway exists, via heat shock protein 90 (HSP90). HSP90 is also known to localize to cilia axonemes, and may act as a signal transductor that interacts with eNOS in the vasculature [62, 63]. HSP90 activation can lead to an increase in eNOS activity while calcium levels are high, and can also lead to more eNOS activity at low calcium levels due to its ability to directly bind to eNOS and increase the binding affinity for calmodulin [64, 65].

4.2 Wound healing

An under researched aspect of NO and primary cilia is their interaction in the vascular smooth muscle cell (VSMC) layer. Depending on blood vessel type, all

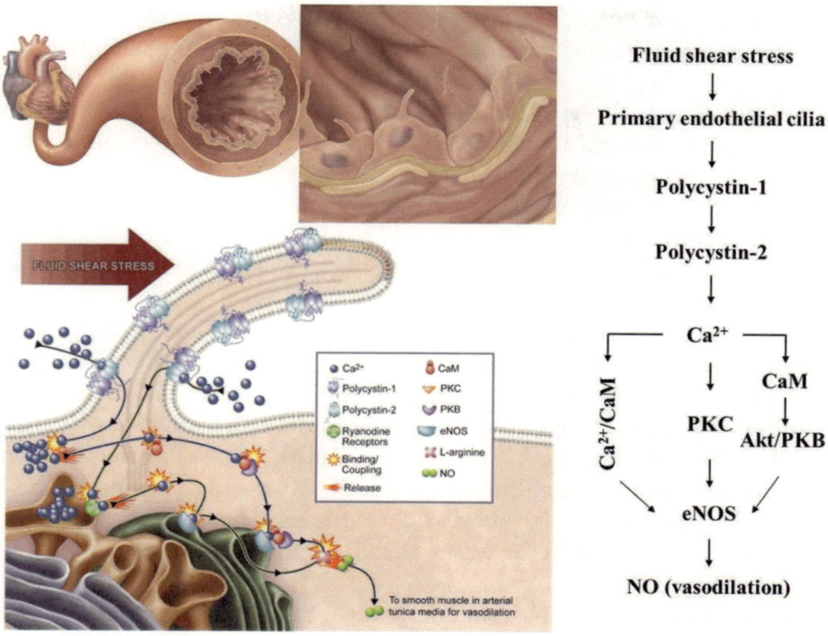

Figure 2.
Primary cilia activation via fluid shear stress and NO signaling in the vascular endothelia. The left panel shows primary cilia bending while under fluid shear stress, with the resultant production and release of nitric oxide (NO). The production and release of NO is dependent on the activation of endothelial primary cilia within the vasculature. The bending of cilia via fluid-shear stress activates the mechanosensory polycystins complex, which initiates the synthesis and the release of NO. This biochemical cascade, shown in the right panel, involves an extracellular calcium influx (Ca^{2+}), followed by the activation of multiple calcium-dependent proteins, including calmodulin (CaM), protein kinase C (PKC) and Akt/PKB. Figure is adopted from Ref. [1].

three isoforms of NOS may exist within the VSMC layers [66]. During normal function, the cilia on the VSMC layer extend towards the extracellular matrix. Under abnormal conditions, such as a scratch wound, VSMC primary cilia will migrate to the wound edge. A recent study showed that VSMC cilia express polycystins, as well as α3-and β1-integrins. When the researchers blocked integrin function, the percent of cilia migrating to the wound edge dropped from about 88% to around 30% [67]. This drop suggests that VSMC primary cilia may be involved in integrin-mediated wound healing.

In further support of the purported role primary cilia play in wound healing, results from experiments using VSMC that lacked cilia showed a slower scratch-wound healing time than the ciliated control cells displayed [67]. While the cilia in the wounded area are directly exposed to constant fluid shear stress from blood flow, activation of the mechanosensory ciliary polycystin complex could occur, resulting in an increase in intracellular calcium [1]. This could potentially trigger vasoconstriction in VSMC, resulting in the isolation of the wounded area to allow the platelets to begin clot formation. As soon as the calcium amount reached a certain level, the calcium/calmodulin complex would form and activate eNOS and nNOS, leading to vasodilation and the next step in the wound healing process.

Further studies by Schneider et al. reveled that, once cellular growth was halted, platelet derived growth factor receptor alpha (PDGFRα), a tyrosine kinase with a prominent function in cellular proliferation, was found to localize to fibroblast primary cilia. Furthermore, ligand activation of PDGFRα will lead to the activation of the AKT and MEK/ERK proliferative pathways. Because AKT and ERK1/2 regulate eNOS activity in endothelial cells, PDGFRα activation could indirectly lead to eNOS activation [68]. Moreover, data from recent studies on endothelial progenitor cells indicate that platelet-derived growth factor AA (PDGF-AA) might contribute a vital role in wound healing, possibly by its effects on angiogenesis through the PI3K/Akt/eNOS signaling pathway [69]. Bone morphological protein (BMP) receptor II (BMPRII), which is highly expressed on endothelial cells in lung vasculature, as well as moderately expressed in smooth muscle, is also involved in cell wound migration. In this pathway, migration is trigged by the ligands BMP2 and BMP4, which result in eNOS being phosphorylated [70]. While not conclusive, this evidence, when taken all together, suggests that primary cilia may have a significant part to play in the wound healing process.

When tissues begin to repair themselves after a wound, clots must be dissolved to maintain proper blood flow. This is known as clot retraction and platelet inhibition. NO is known to inhibit platelet aggregation, secretion, adhesion, and fibrinogen binding; all through activation of guanylyl cyclase and cGMP, alongside the inhibition of thromboxane A2. By this mechanism, platelet aggregation and accumulation are reduced, enabling the clot to dissolve, and the wound to heal fully [71–73]. Given the evidence, it is possible that an interaction between primary cilia and NO could be important in the wound healing and repair processes.

4.3 Dopamine signaling

Hypertension present in polycystic kidney disease (PKD) patients in the later stages of the disease is made worse by increased kidney volume. However, hypertension can also be seen in children, as well as the early stages of PKD, long before renal function starts to deteriorate. Some evidence suggests that an increase in sympathetic activation occurs in these patients, independently of their kidney function. Dopamine, an endogenous neuronal hormone that acts within the sympathetic nervous system, is confirmed to be involved in the regulation of blood pressure. Abnormal dopamine signaling can lead to hypertensive states in humans. Dopamine receptor 1 (D1) and dopamine receptor 5 (D5) receptors have been found to localize to primary cilia [39, 74–76]. While there are no current therapies that target D1 or D5, some studies using dopamine 1-like receptor subtypes showed vasodilatory effects in peripheral arteries [77].

The D5 receptor is thought to have both a chemosensory and mechanosensory role within primary cilia. Subjecting endothelial ciliary knockout cells $pkd1^{-/-}$ (lacking PC-1), and $Tg737^{orpk/orpk}$ cells that have no cilia, to dopamine under static conditions revealed a significantly subdued calcium influx when compared to the control cells. The researchers contributed this to the presence of underdeveloped cilia in the knockout cells, which would have less D5 receptors on them due to their smaller size. Under flow conditions with added dopamine, the mechanosensory function of the cilia knockout cells was restored, in comparison to the untreated knockout cells. Because calcium influx in these cell lines is associated with eNOS activation, the results of this study suggest a potential restoration of lost vasodilatory responses caused by a failed ciliary induction of NO biosynthesis [74]. There is additional evidence that suggests dopamine receptor 2 (D2) may also localize, or possibly get transported to, the primary cilia [78]. In one study, cerebral vasospasms were reversed with dopamine treatment; but when haloperidol, a D2 selective antagonist drug, was administered, the vasorelaxation failed to occur. It was also reported that, after administration of dopamine, a large increase in eNOS and iNOS expression was seen, and administration of haloperidol also blocked this effect [79].

D2 is also possibly transported to the primary cilia under specific conditions to mediate NOS activity within cells. Evidence supporting the role of ciliary dopamine receptors in the mediation of NO can be found in Autosomal Dominant Polycystic

Kidney Disease (ADPKD) patient clinical trials. ADPKD patients experience extra-renal maladies that mainly affect their cardiovascular system, such as hypertension. The hypertensive state could be brought on, in part, by the inability of primary endothelial cilia to respond to alterations in blood pressure. This would cause a failure to synthesize NO. In a study conducted by Lorthioir et al., flow-mediated dilation of normotensive ADPKD patients was compared to that of adults without ADPKD. It was shown that ADPKD patients had significantly less vasodilation during sustained flow increases, as well as a total loss of NO release when compared to those without ADPKD. When ADPKD patients were administered brachial infusions of 0.25–0.5 μg/kg/min of dopamine, there was an increase in flow-mediated dilation, and a statistically significant increase in dilatory response at the highest dose [80]. According to these results, dopamine receptors may facilitate a connection between primary cilia, NO, and blood pressure regulation in ADPKD patients [1].

4.4 Cell proliferation

Primary cilia also help regulate cell proliferation. As stated in the ciliogenesis section, the cilia extends from the basal body, which is composed of mother and daughter centrioles, and cilia are reabsorbed after cell cycle re-entry [12]. In cancerous cell clusters, cilia are missing from the more prolific dividers, which suggests that despite not playing a major role in cell division, primary cilia are important for starting and stopping cell mitosis [81–84].

NO possibly plays a role in cell proliferation as well, in conjunction with primary cilia. NO has been proven to halt the cell cycle by preventing the transition from G1 to S phase, in a dose dependent manner. The spike in NO is caused by an increase in free l-arginine, which is mediated by various cytokines. PC-1 is a known mediator of the JAK/STAT pathway by activating STAT3; when the cytosolic tail of PC-1 is cleaved upon once luminal flow halts, it can coactivate STAT-1, −3, and −6, along with JAK2. The PC-1 tail causes the cells to sensitize to cytokines and growth factor signaling, which then causes an exaggerated cellular response, which could potentially lead to an increase in l-arginine [85, 86]. Through this mechanism, overly prolific cell division would be arrested.

The superfamily of TGF-β signaling provides a fascinating system of cellular crosstalk, in which the effects of the same ligand can be unique depending on the cell type and the physiological conditions. This family is composed of more than 30 different ligand types of the TGF-β-activin-Nodal BMP subfamilies that can activate receptor serine/threonine kinases of types I and II (TGFβRI/II and BMP-RI/II, respectively). Ciliopathies widely overlap with phenotypes associated with aberrant TGF-β/BMP signaling. Prominent examples include structural heart defects associated with congenital heart disease (CHD) [87], suggesting that cardiac primary cilia may contribute to cellular events regulated by TGF-β/BMP signaling events during heart development. Moreover, different components of the TGF-β signalosome, including TGF-βRI, TGF-βRII, SMAD2/3, SMAD4, and SMAD7 are present at the cilia-centrosome axis. In a recent study, Feng et al. concluded that high salt (HS)-induced endothelial dysfunction and the development of salt-dependent increases in blood pressure (BP) were related to endothelial TGF-β signaling. Specifically, TGF-β-dependent ALK5 signaling increases endothelial NADPH oxidase-4 (NOX4), an enzyme that produces hydrogen peroxide, which limits NO bioavailability and ultimately promotes increased BP [88]. BMPRII contributes to cell proliferation through its interactions with primary cilia, eNOS, and NO. Using pulmonary artery endothelial cells, studies have shown that stimulation of BMPRII results in eNOS activation. BMPRII ligands BMP2 and BMP4 stimulate eNOS phosphorylation at a regulatory site via activation by protein kinase A. This eNOS stimulation results in increased NO bioavailability; loss of BMPRII function, therefore, is proposed to contribute to endothelial dysfunction [70].

5. Conclusion

Primary cilia and nitric oxide are both essential for normal tissue functions. While they can operate independently of one another, often their roles and cellular pathways are complementary. While this chapter only touched on a small portion of their possible connections, scattered research suggests a more complex linkage between the two in many more organ systems and cellular pathways. However, current research on the direct links between primary cilia and NO are scarce. The aim of this chapter was to discuss the more well-known links between primary cilia and NO, and to initiate discussion leading to further examination of the topics covered. Revealing the connections between cilia and NO would provide insight into various ciliopathies and could reveal new targets for therapies.

Author details

Sidney T. Ley and Wissam A. AbouAlaiwi*
Department of Pharmacology and Experimental Therapeutics, The University of Toledo, USA

*Address all correspondence to: wissam.abou-alaiwi@utoledo.edu

IntechOpen

© 2019 The Author(s). Licensee IntechOpen. This chapter is distributed under the terms of the Creative Commons Attribution License (http://creativecommons.org/licenses/by/3.0), which permits unrestricted use, distribution, and reproduction in any medium, provided the original work is properly cited.

References

[1] Saternos HC, AbouAlaiwi WA. Signaling interplay between primary cilia and nitric oxide: A mini review. Nitric Oxide-Biology and Chemistry. 2018;**80**:108-112

[2] Waters AM, Beales PL. Ciliopathies: An expanding disease spectrum. Pediatric Nephrology. 2011;**26**(7):1039-1056

[3] Nauli SM et al. Endothelial cilia are fluid shear sensors that regulate calcium signaling and nitric oxide production through polycystin-1. Circulation. 2008;**117**(9):1161-1171

[4] AbouAlaiwi WA et al. Ciliary polycystin-2 is a mechanosensitive calcium channel involved in nitric oxide signaling cascades. Circulation Research. 2009;**104**(7):860-869

[5] Mitchison HM, Valente EM. Motile and non-motile cilia in human pathology: From function to phenotypes. The Journal of Pathology. 2017;**241**(2):294-309

[6] Brooks ER, Wallingford JB. Multiciliated cells. Current Biology: CB. 2014;**24**(19):R973-R982

[7] Omran AJA et al. Alcohol consumption impairs the ependymal cilia motility in the brain ventricles. Scientific Reports. 2017;**7**(1):13652-13652

[8] Nonaka S et al. Determination of left–right patterning of the mouse embryo by artificial nodal flow. Nature. 2002;**418**(6893):96

[9] Veland IR et al. Primary cilia and signaling pathways in mammalian development, health and disease. Nephron. Physiology. 2009;**111**(3):p39-p53

[10] Arnaiz O et al. Remodeling Cildb, a popular database for cilia and links for ciliopathies. Cilia. 2014;**3**(1):9

[11] Pazour GJ et al. Proteomic analysis of a eukaryotic cilium. The Journal of Cell Biology. 2005;**170**(1):103-113

[12] Avasthi P, Marshall WF. Stages of ciliogenesis and regulation of ciliary length. Differentiation. 2012;**83**(2):S30-S42

[13] Graser S et al. Cep164, a novel centriole appendage protein required for primary cilium formation. The Journal of Cell Biology. 2007;**179**(2):321-330

[14] Kim S et al. Nde1-mediated inhibition of ciliogenesis affects cell cycle re-entry. Nature Cell Biology. 2011;**13**(4):351

[15] Korobeynikov V, Deneka AY, Golemis EA. Mechanisms for nonmitotic activation of Aurora-A at cilia. Biochemical Society Transactions. 2017;**45**(1):37-49

[16] Goetz SC, Anderson KV. The primary cilium: A signalling centre during vertebrate development. Nature Reviews Genetics. 2010;**11**(5):331

[17] Reiter JF, Blacque OE, Leroux MR. The base of the cilium: Roles for transition fibres and the transition zone in ciliary formation, maintenance and compartmentalization. EMBO Reports. 2012;**13**(7):608-618

[18] Lin H, Guo S, Dutcher SK. RPGRIP1L helps to establish the ciliary gate for entry of proteins. Journal of Cell Science. 2018;**131**(20):jcs220905

[19] Fisch C, Dupuis-Williams P. Ultrastructure of cilia and flagella–back to the future! Biology of the Cell. 2011;**103**(6):249-270

[20] Deane JA et al. Localization of intraflagellar transport protein IFT52 identifies basal body transitional

fibers as the docking site for IFT particles. Current Biology. 2001;**11**(20):1586-1590

[21] Pedersen LB, Rosenbaum JL. Chapter two intraflagellar transport (IFT): Role in ciliary assembly, resorption and signalling. Current Topics in Developmental Biology. 2008;**85**:23-61

[22] Taschner M, Lorentzen E. The intraflagellar transport machinery. Cold Spring Harbor Perspectives in Biology. 2016;**8**(10):a028092

[23] Mizukami I, Gall J. Centriole replication: II. Sperm formation in the fern, Marsilea, and the cycad, Zamia. The Journal of Cell Biology. 1966;**29**(1):97-111

[24] Hodges ME et al. The evolution of land plant cilia. New Phytologist. 2012;**195**(3):526-540

[25] Nielsen C. Structure and function of metazoan ciliary bands and their phylogenetic significance. Acta Zoologica. 1987;**68**(4):205-262

[26] Lidow MS, Menco BPM. Observations on axonemes and membranes of olfactory and respiratory cilia in frogs and rats using tannic acid-supplemented fixation and photographic rotation. Journal of Ultrastructure Research. 1984;**86**(1):18-30

[27] Satir P, Christensen ST. Overview of structure and function of mammalian cilia. Annual Review of Physiology. 2007;**69**: 377-400

[28] King SM. Axonemal dynein arms. Cold Spring Harbor Perspectives in Biology. 2016;**8**(11):a028100

[29] Golestanian R, Yeomans JM, Uchida N. Hydrodynamic synchronization at low Reynolds number. Soft Matter. 2011;7(7):3074-3082

[30] Machemer H. Ciliary activity and the origin of metachrony in Paramecium: Effects of increased viscosity. Journal of Experimental Biology. 1972;**57**(1):239-259

[31] Satir P, Sleigh MA. The physiology of cilia and mucociliary interactions. Annual Review of Physiology. 1990;**52**(1):137-155

[32] Elgeti J, Gompper G. Emergence of metachronal waves in cilia arrays. Proceedings of the National Academy of Sciences. 2013;**110**(12):4470-4475

[33] Davenport JR, Yoder BK. An incredible decade for the primary cilium: A look at a once-forgotten organelle. American Journal of Physiology. Renal Physiology. 2005;**289**(6):F1159-F1169

[34] Chen JC, Jacobs CR. Cellular and molecular mechanotransduction in bone. In: Osteoporosis: Fourth Edition. Amsterdam: Elsevier Inc.; 2013. pp. 453-475

[35] Gonçalves J, Pelletier L. The ciliary transition zone: Finding the pieces and assembling the gate. Molecules and Cells. 2017;**40**(4):243-253

[36] Lechtreck KF. IFT–cargo interactions and protein transport in cilia. Trends in Biochemical Sciences. 2015;**40**(12):765-778

[37] Hu Q, Nelson WJ. Ciliary diffusion barrier: The gatekeeper for the primary cilium compartment. Cytoskeleton. 2011;**68**(6):313-324

[38] Malicki J, Avidor-Reiss T. From the cytoplasm into the cilium: Bon voyage. Organogenesis. 2014;**10**(1):138-157

[39] Leaf A, Von Zastrow M. Dopamine receptors reveal an essential role of IFT-B, KIF17, and Rab23 in delivering specific receptors to primary cilia. eLife. 2015;**4**:e06996

[40] Mukhopadhyay S et al. Trafficking to the primary cilium membrane. Molecular Biology of the Cell. 2017;**28**(2):233-239

[41] Kim H et al. Ciliary membrane proteins traffic through the Golgi via a Rabep1/GGA1/Arl3-dependent mechanism. Nature Communications. 2014;**5**:5482

[42] Szymanska K, Johnson CA. The transition zone: An essential functional compartment of cilia. Cilia. 2012;**1**(1):10

[43] Klink BU et al. A recombinant BBSome core complex and how it interacts with ciliary cargo. eLife. 2017;**6**:e27434

[44] Zhang Q et al. BBS7 is required for BBSome formation and its absence in mice results in Bardet-Biedl syndrome phenotypes and selective abnormalities in membrane protein trafficking. Journal of Cell Science. 2013;**126**(11):2372-2380

[45] Jensen VL et al. Whole-organism developmental expression profiling identifies RAB-28 as a novel ciliary GTPase associated with the BBSome and intraflagellar transport. PLoS Genetics. 2016;**12**(12):e1006469

[46] Geng L et al. Polycystin-2 traffics to cilia independently of polycystin-1 by using an N-terminal RVxP motif. Journal of Cell Science. 2006;**119**(7):1383-1395

[47] Leyssac PP. Changes in single nephron renin release are mediated by tubular fluid flow rate. Kidney International. 1986;**30**(3):332-339

[48] Praetorius HA, Spring KR. The renal cell primary cilium functions as a flow sensor. Current Opinion in Nephrology and Hypertension. 2003;**12**(5):517-520

[49] Praetorius H, Spring KR. Bending the MDCK cell primary cilium increases intracellular calcium. The Journal of Membrane Biology. 2001;**184**(1):71-79

[50] Piperi C, Basdra EK. Polycystins and mechanotransduction: From physiology to disease. World Journal of Experimental Medicine. 2015;**5**(4):200

[51] Yoder BK, Hou X, Guay-Woodford LM. The polycystic kidney disease proteins, polycystin-1, polycystin-2, polaris, and cystin, are co-localized in renal cilia. Journal of the American Society of Nephrology. 2002;**13**(10):2508-2516

[52] Trouillon R. Biological applications of the electrochemical sensing of nitric oxide: Fundamentals and recent developments. Biological Chemistry. 2013;**394**(1):17-33

[53] Wu G, Morris SM Jr. Arginine metabolism: Nitric oxide and beyond. The Biochemical Journal. 1998;**336** (Pt 1):1-17

[54] Luiking YC, Engelen MP, Deutz NE. Regulation of nitric oxide production in health and disease. Current Opinion in Clinical Nutrition and Metabolic Care. 2010;**13**(1):97-104

[55] Forstermann U, Sessa WC. Nitric oxide synthases: Regulation and function. European Heart Journal. 2012;**33**(7):829-837

[56] Boo YC, Jo H. Flow-dependent regulation of endothelial nitric oxide synthase: Role of protein kinases. American Journal of Physiology-Cell Physiology. 2003;**285**(3):C499-C508

[57] Su S et al. Genetically encoded calcium indicator illuminates calcium dynamics in primary cilia. Nature Methods. 2013;**10**(11):1105

[58] Nauli SM, Pala R, Kleene SJ. Calcium channels in primary cilia. Current Opinion in Nephrology and Hypertension. 2016;**25**(5):452

[59] Delling M et al. Primary cilia are not calcium-responsive mechanosensors. Nature. 2016;**531**(7596):656

[60] Leonard PM et al. Crystal structure of the Lrp-like transcriptional regulator from the archaeon *Pyrococcus furiosus*. The EMBO Journal. 2001;**20**(5):990-997

[61] Stahmann N et al. Activation of AMP-activated protein kinase by vascular endothelial growth factor mediates endothelial angiogenesis independently of nitric-oxide synthase. Journal of Biological Chemistry. 2010;**285**(14):10638-10652

[62] Chen Y et al. Differential effects of heat shock protein 90 and serine 1179 phosphorylation on endothelial nitric oxide synthase activity and on its cofactors. PLoS One. 2017;**12**(6):e0179978

[63] Chen K, Pittman RN, Popel AS. Nitric oxide in the vasculature: Where does it come from and where does it go? A quantitative perspective. Antioxidants & Redox Signaling. 2008;**10**(7):1185-1198

[64] Takahashi S, Mendelsohn ME. Synergistic activation of endothelial nitric-oxide synthase (eNOS) by HSP90 and Akt calcium-independent eNOS activation involves formation of an HSP90-Akt-CaM-bound eNOS complex. Journal of Biological Chemistry. 2003;**278**(33):30821-30827

[65] Takahashi S, Mendelsohn ME. Calmodulin-dependent and-independent activation of endothelial nitric-oxide synthase by heat shock protein 90. Journal of Biological Chemistry. 2003;**278**(11):9339-9344

[66] Buchwalow IB et al. Vascular smooth muscle and nitric oxide synthase. The FASEB Journal. 2002;**16**(6):500-508

[67] Lu C et al. Non-random distribution and sensory functions of primary cilia in vascular smooth muscle cells. Kidney and Blood Pressure Research. 2008;**31**(3):171-184

[68] Schneider L et al. PDGFRαα signaling is regulated through the primary cilium in fibroblasts. Current Biology. 2005;**15**(20):1861-1866

[69] Wu LW et al. Platelet-derived growth factor-AA is a substantial factor in the ability of adipose-derived stem cells and endothelial progenitor cells to enhance wound healing. The FASEB Journal. 2019;**33**(2):2388-2395

[70] Gangopahyay A et al. Bone morphogenetic protein receptor II is a novel mediator of endothelial nitric-oxide synthase activation. The Journal of Biological Chemistry. 2011;**286**(38):33134-33140

[71] Wang G-R et al. Mechanism of platelet inhibition by nitric oxide: in vivo phosphorylation of thromboxane receptor by cyclic GMP-dependent protein kinase. Proceedings of the National Academy of Sciences. 1998;**95**(9):4888-4893

[72] Du X. A new mechanism for nitric oxide–and cGMP-mediated platelet inhibition. Blood. 2007;**109**(2):392-393

[73] Riddell DR, Owen JS. Nitric oxide and platelet aggregation. In: Vitamins & Hormones. Amsterdam: Elsevier; 1997. pp. 25-48

[74] Abdul-Majeed S, Nauli SM. Dopamine receptor type 5 in the primary cilia has dual chemo-and mechano-sensory roles. Hypertension. 2011;**58**(2):325-331

[75] Upadhyay V et al. Roles of dopamine receptor on chemosensory and mechanosensory primary cilia in renal epithelial cells. Frontiers in Physiology. 2014;**5**:72

[76] Marley A, von Zastrow M. DISC1 regulates primary cilia that display specific dopamine receptors. PLoS One. 2010;**5**(5):e10902

[77] Asghar M et al. Potential dopamine-1 receptor stimulation in hypertension management. Current Hypertension Reports. 2011;**13**(4):294-302

[78] Omori Y et al. Identification of G protein-coupled receptors (GPCRs) in primary cilia and their possible involvement in body weight control. PLoS One. 2015;**10**(6):e0128422

[79] Pyne-Geithman GJ et al. Dopamine D 2-receptor-mediated increase in vascular and endothelial NOS activity ameliorates cerebral vasospasm after subarachnoid hemorrhage in vitro. Neurocritical Care. 2009;**10**(2):225

[80] Lorthioir A et al. Polycystin deficiency induces dopamine-reversible alterations in flow-mediated dilatation and vascular nitric oxide release in humans. Kidney International. 2015;**87**(2):465-472

[81] Goto H, Inoko A, Inagaki M. Cell cycle progression by the repression of primary cilia formation in proliferating cells. Cellular and Molecular Life Sciences. 2013;**70**(20):3893-3905

[82] Ke Y-N, Yang W-X. Primary cilium: An elaborate structure that blocks cell division? Gene. 2014;**547**(2):175-185

[83] Plotnikova OV, Golemis EA, Pugacheva EN. Cell cycle–dependent ciliogenesis and cancer. Cancer Research. 2008;**68**(7):2058-2061

[84] Cao M, Zhong Q. Cilia in autophagy and cancer. Cilia. 2015;**5**(1):4

[85] Talbot JJ et al. Polycystin-1 regulates STAT activity by a dual mechanism. Proceedings of the National Academy of Sciences. 2011;**108**(19):7985-7990

[86] Weimbs T, Olsan EE, Talbot JJ. Regulation of STATs by polycystin-1 and their role in polycystic kidney disease. Jak-Stat. 2013;**2**(2):e23650

[87] Koefoed K et al. Cilia and coordination of signaling networks during heart development. Organogenesis. 2014;**10**(1):108-125

[88] Feng W et al. Transforming growth factor-beta mediates endothelial dysfunction in rats during high salt intake. American Journal of Physiology. Renal Physiology. 2015;**309**(12):F1018-F1025

Chapter 4

Endogenous and Inhaled Nitric Oxide for the Treatment of Pulmonary Hypertension

Kazuo Maruyama, Junko Maruyama and Hirofumi Sawada

Abstract

Since the discovery of nitric oxide (NO) as a physiological substance produced in the endothelium, the impairment of endothelial NO production and reactivity of the pulmonary vasculature to NO have been described in animal models and patients with pulmonary hypertension (PH). The NO synthase-NO-cyclic guanosine monophosphate (cGMP) pathway is impaired in pulmonary arterial hypertension (PAH), pulmonary veno-occlusive disease (PVOD), pulmonary capillary hemangiomatosis (PCH), chronic obstructive pulmonary disease (COPD), and idiopathic pulmonary fibrosis (IPF). Pioneering clinicians conceived that NO can be administered to the lung by inhalation and used this strategy to treat PH in humans and acute hypoxic PH in animal models. Inhaled NO (iNO) selectively decreases pulmonary arterial pressure with no changes in systemic arterial pressure. When iNO diffuses into the blood, it is converted to NO_3^-, thereby losing its vasodilatory effects. NO might then be recycled in hypoxic remote organs, where NO_3^- and NO_2^- are reduced to NO. In the present chapter, the metabolic fate of iNO, based on previous air pollution research in Japan, is discussed. Then, we describe recent clinical applications of iNO in pediatric patients with various diseases, including bronchopulmonary dysplasia (BPD), persistent PH of neonates, and congenital diaphragmatic hernia (CDH). We also summarize the role of iNO in the catheterization lab, including acute vasoreactivity testing to assess prognosis, indications for specific PH therapy, and operability of congenital heart disease.

Keywords: nitric oxide inhalation, pulmonary hypertension, metabolism of nitric oxide, nitrate, pediatric, neonate

1. Introduction

In 1980, Furchgott noted that the endothelium produces and releases a vasodilatory substance named endothelium-derived relaxing factor (EDRF), which diffuses into adjacent vascular smooth muscle cells and results in vascular relaxation [1]. In 1987, EDRF was identified as nitric oxide (NO) by Ignarro [2] and Moncada [3]. Murad reported the vasodilatory effect of nitroglycerin and NO formation from nitroglycerin in 1977 [4]. However, at the time, it was not known that endogenous NO is produced and released as a physiological substance in the body, especially in the vascular endothelium.

High-temperature combustion accelerates the reaction of oxygen and nitrogen in air to generate nitrogen oxides (NOx), such as NO, NO_2, and N_2O_3. A common source of NOx is car engines, among which diesel engines have particularly high production. NO reacts with O_2 to produce NO_2, which is more toxic than NO. Thus, NOx, including NO, is considered an air pollutant. Accordingly, measuring instruments and NO gas standards with known concentrations are needed to assess NO concentrations in air. In addition, NO gas has various industrial applications, including uses in the production of chemicals, semiconductors, integrated circuits, and memory storage elements and devices. Therefore, measuring instruments for NO and the delivery of NO from gas cylinders were developed long before the discovery of NO as a physiological substance in the body. NO is now recognized as a gas and a physiological substance. Pioneering clinicians determined that "as a gas, NO can be administered to the body through the lung." It was fortunate for these clinicians who first conducted NO inhalation in humans that measuring instruments for NO and NO cylinders were available.

The present chapter discusses endogenous NO production in normal and hypertensive pulmonary vasculature, the history of NO inhalation for therapeutic use, the fate of inhaled NO (iNO), effects of iNO in remote organs other than the lung, and iNO as a therapeutic strategy in pediatrics.

2. Endogenous NO and its role in the pathogenesis and pathophysiology of pulmonary hypertension

NO is primarily synthesized by endothelial NO synthase (eNOS, NOSIII) in pulmonary vascular endothelial cells. NO reacts with a receptor, soluble guanylate cyclase (sGC), in adjacent smooth muscle cells. Activated sGC produces cGMP, which stimulates protein kinase G (PKG) and exerts many physiological effects, including pulmonary vascular relaxation. The inhibition of NO production by L-NMMA (*N*-omega-monomethyl-L-arginine, a NOS inhibitor) decreases pulmonary flow in conscious healthy adults [5]. A deficiency in eNOS, but not iNOS or neuronal NOS, induces augmented hypoxic pulmonary vasoconstriction and a lack of endothelium-dependent vasodilation [6]. These findings support the important roles of the eNOS-NO-cyclic guanosine monophosphate (GMP) pathway in maintaining pulmonary circulation. Alteration of eNOS expression and/or function may contribute to decreased NO synthesis in pulmonary hypertension (PH). Human PH has many different etiologies. Depending on the pathological state, patients may exhibit alterations in the eNOS-NO-cGMP pathway.

2.1 Effects of NO in isolated pulmonary arteries

The effects of NO differ among cell types. NO induces relaxation in vascular smooth muscle cells, prevents aggregation and adhesion in platelets, prevents adhesion in leucocytes, and acts as a neurotransmitter in synapses. Thus, NO regulates various cell functions. In the vasculature, NO is released from endothelial cells, reaches adjacent smooth muscle cells, and causes vascular relaxation, indicating that it functions in intercellular signaling. Among the physiological roles of endogenous NO, vascular relaxation was discovered first.

In isolated rat main pulmonary arterial rings precontracted with prostaglandin $F_{2\alpha}$ (PGF_{2a}), acetylcholine (ACh) induces relaxation in endothelium-preserved pulmonary arteries, but not in endothelium-denuded pulmonary arteries, suggesting that the endothelium in pulmonary arteries produces a relaxation-inducing substance in response to acetylcholine (**Figure 1(a)** and (**b**)).

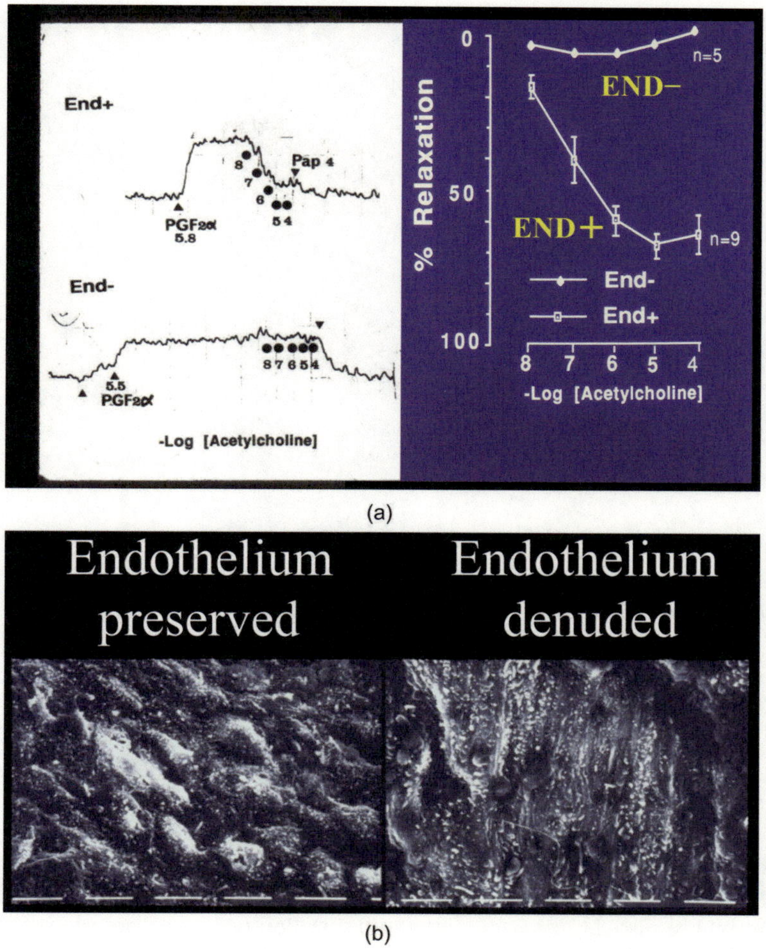

Figure 1.
(a) Acetylcholine induces relaxation in endothelium-preserved pulmonary arterial rings. Pulmonary artery rings were obtained from normal control air rats. Rings were suspended in an 20 ml organ bath, and isometric tension was measured. Relaxation responses to acetylcholine (Ach) in endothelium-preserved (END+) and endothelium-denuded (END-) rings of the extrapulmonary artery were obtained. Endothelium was removed by gently rubbing luminal surface by fine stainless wire in endothelium-denuded rings. Rings were precontracted with prostaglandin F_{2a} (PGF_{2a}). Relaxation induced by 10^{-4} M papaverine (Pap 4) was taken as 100%. Bars mean standard error. Relaxation responses to ACh were abolished in the endothelium-denuded pulmonary arterial rings, showing that pulmonary vascular endothelium releases vasorelaxation substance named endothelium-derived relaxing factor (EDRF). The absence of the endothelium was confirmed by scanning electron micrography (b). END−, endothelium-denuded rings; END+, endothelium-preserved rings; 8, 10^{-8} mol/L, the same for 7, 6, 5, and 4. (B) Scanning electron micrograph of the endothelium-preserved pulmonary artery and endothelium-denuded pulmonary artery. Endothelium was removed by gently rubbing luminal surface by fine stainless wire. Left: luminal surface of the endothelium-preserved pulmonary artery. Right: luminal surface of the endothelium-denuded pulmonary artery.

In pulmonary arteries isolated from experimental PH models (chronic hypoxic PH in rat), the relaxation response to acetylcholine (ACh) is depressed, as observed in endothelium-denuded arteries, suggesting that endothelial function is impaired in PH arteries. Both ACh- and sodium nitroprusside (SNP, an NO donor)-induced relaxations were impaired in pulmonary arteries from rats with chronic hypoxic PH, suggesting that NO-induced relaxation is depressed in hypertensive pulmonary arteries [7, 8] (**Figure 2**). However, the magnitude of impairment seems to be higher in ACh-induced endothelium-dependent relaxation than in SNP-induced endothelium-independent relaxation [7].

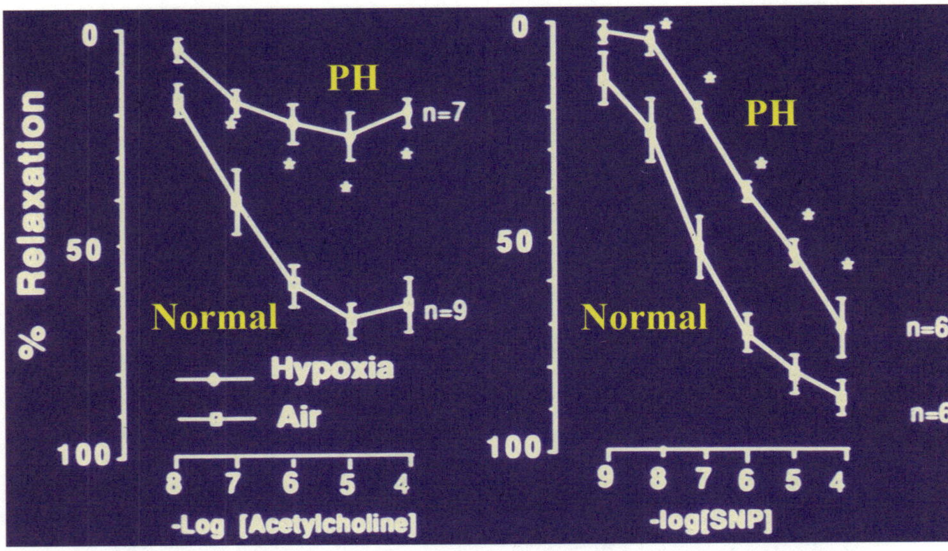

Figure 2.
The relaxation responses are depressed in isolated pulmonary arterial rings from chronic hypoxic pulmonary hypertensive rat. Pulmonary artery rings were obtained from normal control air rats and rats exposed to 10 days of hypoxia with chronic hypoxic pulmonary hypertension (PH). Isometric tension was measured. Relaxation responses to acetylcholine (ACh) in prostaglandin F_{2a} (PGF_{2a})-precontracted rings of extrapulmonary were recorded. Relaxation induced by 10^{-4} M papaverine (Pap 4) was taken as 100%. Relaxation responses to ACh were depressed in rings from rats with chronic hypoxic PH, showing that the release of vasorelaxation substance is impaired in PH rings. Although the relaxation responses to sodium nitroprusside (SNP) are impaired in pH rings compared with control, this means that there was a room where SNP could cause relaxation in PH rings from chronic hypoxic PH.

The relaxation responses to SNP are caused by the liberation of NO from SNP. To determine the vasodilatory effects of NO directly, a NO solution was made by bubbling 10% NO in pure N_2 into deoxygenated distilled water. Although depressed, the relaxation responses were indeed induced by NO in hypertensive pulmonary arteries [7] (**Figure 3**). Importantly, iNO exhibits selectivity, resulting in vasodilation in pulmonary arteries (**Figure 4**) when administered by inhalation through the trachea. The intravenous injection of NO donors simultaneously decreases both pulmonary and systemic arterial pressure.

2.2 Pulmonary arterial hypertension

2.2.1 eNOS in pulmonary arterial hypertension in humans

Although the role of the eNOS-derived NO-related pathway in pulmonary arterial hypertension (PAH) has been determined, its pathophysiology remains unclear. eNOS plays a key role in this pathway. Giaid et al. detected decreased eNOS protein expression in human lungs with PAH [9]. Additionally, exhaled NO has been found to be lower in patients with PAH than in controls [10]. Subsequent studies have reported increased eNOS protein expression in plexiform lesions in PAH [11] and increased eNOS activity in idiopathic PAH (IPAH) lungs, despite no change in NOS expression [12]. The membrane protein caveolin-1 (CAV1) is a crucial negative regulator of eNOS activity. The CAV1 expression is decreased in IPAH lungs, which might lead to persistent eNOS activation, the accumulation of dysfunctional (i.e., uncoupled) eNOS, the formation of peroxynitrite, and the impairment of PKG

DOI: http://dx.doi.org/10.5772/intechopen.89381

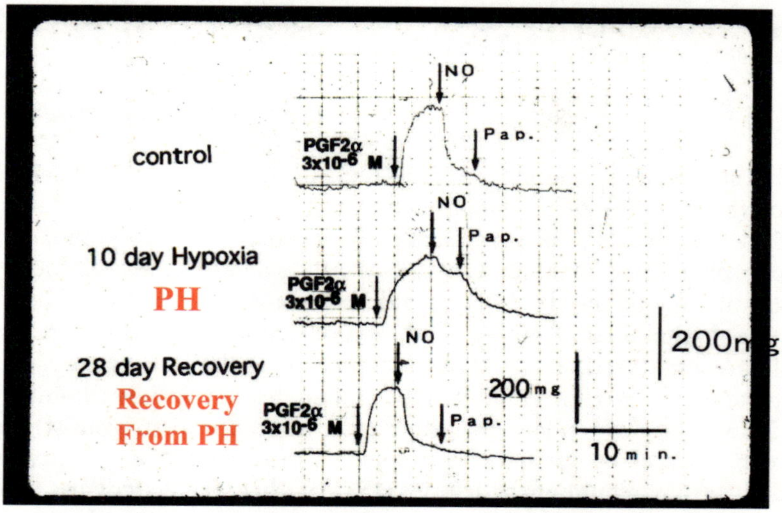

Figure 3.
NO solution (0.16–0.2 mM NO) caused the relaxation responses in isolated normal and pulmonary hypertensive arterial rings. Pulmonary artery rings were obtained from normal control air rats (A), rats exposed to 10 days of hypoxia with chronic hypoxic pulmonary hypertension (PH) (B), and rats after 28 days of recovery in room air from chronic hypoxia (C). NO solution was made by bubbling 10% NO through deoxygenated distilled water, which results in 0.16–0.2 mM concentration. Aliquots (0.5 ml) of this solution were applied to the organ bath. Papaverine (Pap) was introduced to obtain maximal relaxation. Relaxation responses to NO solution to prostaglandin F2a-precontracted rings were recorded. (A) NO-induced relaxation in pulmonary artery rings from normal rats. (B) Response to NO was depressed in pulmonary artery rings from chronic hypoxic rats. (C) The relaxation response returned to normal after 28 days of recovery from chronic hypoxic pulmonary hypertension. The result of (B) showed that NO could dilate hypertensive pulmonary vascular smooth muscles, although depressed compared to normal.

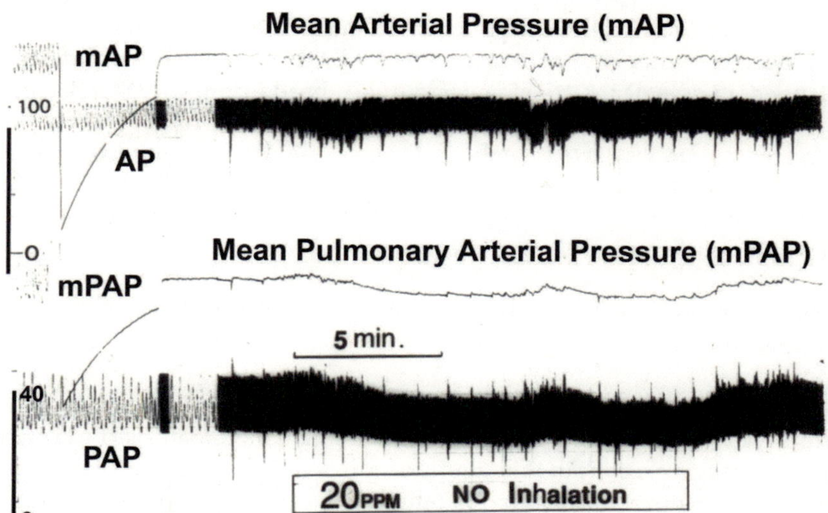

Figure 4.
Inhaled NO as selective pulmonary vasodilator in pulmonary hypertensive rats. A pulmonary artery catheter was introduced via the right external jugular vein into the pulmonary artery by a closed chest technique. Pulmonary arterial pressure(PAP) was recorded with rat fully awake in a pulmonary hypertensive rat (19 days after the single injection of monocrotaline). About 20 ppm NO inhalation decreased PAP with no change of arterial pressure. When NO inhalation was discontinued, the PAP returned to baseline.

kinase activity via tyrosine nitration [12]. Increased eNOS activity and/or expression might not be associated with NO production in PAH.

2.2.2 eNOS-NO pathway in animal models of pulmonary arterial hypertension

Animal studies using a monocrotaline (MCT)-induced PAH rat model, which is characterized by pulmonary endothelial damage and perivascular inflammation in the early pathological stage, have shown decreased eNOS expression [13] and/or phosphorylated eNOS activity [14] as well as decreases in sGC and PKG. Vasodilation induced by ACh, an endothelium-dependent NO-related vasodilator, was also impaired. Another adult rat model of severe PAH with precapillary obliterative lesion (SU/Hx model) shows similarities in the pulmonary vascular pathology to that of PAH in adults. This SU/Hx model, induced by combined SUGEN5416 (a vascular endothelial growth factor receptor II antagonist) and exposure to chronic hypoxia, showed a reduction of ACh-induced NO production and/or release in pulmonary arteries [15]. Another recent study has reported decreased CAV1 expression in the same model [16]. eNOS also translocates from cell surface caveolae to cytoplasmic and perinuclear regions in pathological state [17]. Consequently, the amount of NO production is decreased. Accordingly, the pathogenesis and progression of PAH may be partially induced by endothelial dysfunction associated with suppression of the eNOS-NO-related pathway.

2.2.3 eNOS-NO-cGMP pathway and BMPRII

Genetic variants in the eNOS-NO-cGMP pathway might cause PAH. CAV1 plays an important role in NO signaling in PAH. Mutations in CAV1 have been identified in PAH [18]. The gene encoding bone morphogenetic protein receptor 2 (BMPRII) is frequently mutated in heritable PAH [19, 20] and adult IPAH [19, 21]. BMPRII is a member of the transforming growth factor (TGF)-β receptor superfamily, localized to caveolae, and interacts with CAV1 in vascular smooth muscle cells [22, 23]. Recent studies have demonstrated that BMPRII deficiency promotes SRC-dependent caveolar trafficking defects [24]. In addition, CAV1-deficient mice have shown reduced BMPRII expression after exposure to chronic hypoxia [16]. In MCT-treated pulmonary arterial endothelial cells, BMPRII was increasingly trapped intracellularly together with increased trapping CAV1 and eNOS. These results suggest that NO-cGMP-related dysfunction and BMPRII deficiency are closely related to and play a significant role in the pathogenesis of PAH.

2.3 Pulmonary veno-occlusive disease and pulmonary capillary hemangiomatosis

Pulmonary veno-occlusive disease (PVOD), classified as a PAH subgroup, is inextricably associated with pulmonary capillary hemangiomatosis (PCH) [25]. Pulmonary vascular lesions in this condition are mainly detected in postcapillary venules and veins but are also found in pulmonary capillaries and arteries [25]. The pathogenesis is heterogeneous and poorly understood [25]. These are rare diseases, and few studies have focused on the pathogenesis and pathophysiology. Kradin et al. reported that eNOS expression in abnormal capillary lesions is significantly decreased in patients with PCH with pulmonary vascular remodeling and concomitant pulmonary hypertension and is minimally decreased or not decreased in patients without pulmonary vascular remodeling [26]. These results suggest that the alteration of eNOS expression is associated with the pathogenesis of these complicated conditions. Further experiments are necessary to determine the precise role of the eNOS-NO-cGMP pathway in PVOD/PCH.

Biallelic mutations in eukaryotic translation initiation factor 2α kinase 4 (*EIF2AK4*) have been identified in familial and idiopathic PVOD/PCH. *EIF2AK4* encodes general control nonderepressible 2 (GCN2) [27]. The most common experimental models of these conditions are mitomycin C (MMC)-treated rats and mice [28]. Interestingly, MMC dose dependently induces pulmonary GCN2 depletion [28]. *EIF2AK4* mutations are also found in sporadic PVOD/PCH [27]. Mutation carriers have distinct histological features, including strong muscular hyperplasia of the interlobular septal vein as well as arterial severe intimal fibrosis [29]. *EIF2AK4* is activated by amino acid depletion. Because L-arginine, a substrate of NOS, is depleted during NO production, *EIF2AK4* activation can be induced by eNOS activity associated with L-arginine depletion.

2.4 Lung disease-related and/or alveolar hypoxia-induced pH

The pathophysiologic features of lung diseases include chronic obstructive pulmonary disease (COPD) and idiopathic pulmonary fibrosis (IPF) and mixed pathologic diseases, including combined pulmonary fibrosis and emphysema. All involve alveolar hypoxia and subsequent hypoxic pulmonary vasoconstriction. eNOS expression is upregulated in acute hypoxia in rat lungs [30]. eNOS expression increases in a time-dependent manner in rats during the development of hypoxia-induced PH [31–33], while eNOS activity is impaired [34]. The production of tetrahydrobiopterin (BH_4), an obligatory cofactor for generating the active dimer form of eNOS, was altered in hypoxic conditions [34]. An imbalance between BH_4 and dihydrobiopterin (BH_2) may cause eNOS uncoupling and inactive monomer formation [34]. Several studies have reported decreased eNOS expression and/or activity in patients with COPD [35, 36], with severity of endothelial dysfunction correlated with degree of airflow obstruction [36]. eNOS is also absent in pulmonary arteries of patients with IPF [37]. As the histological features of this disease differ from those of COPD, the pathogenesis of IPF-induced PH may include a multifactorial and complex process involving proinflammatory cytokines and growth factors.

3. Inhaled nitric oxide

3.1 Inhaled NO as a selective pulmonary vasodilator

In 1988, at the international conference of the American Thoracic Society, Higenbottam presented his team's paper titled "Inhaled endothelium-derived relaxing factor (EDRF) in primary pulmonary hypertension (PPH)," including the first description of NO inhalation in humans with pulmonary arterial hypertension for laboratory use [38]. In 1991, Lancet published the study [39], showing that 40 ppm NO inhalation selectively reduces PAP, with no changes in systemic pressure. Frostell et al. also showed that inhaled NO (5–80 ppm) causes selective pulmonary arterial dilatation, without changes in systemic arterial pressure in sheep [40], where PAP elevation was induced by hypoxic pulmonary vasoconstriction. Both research groups referenced the studies by Yoshida, Kasama, and Kitabatake about the metabolic fate of iNO [41, 42] because toxicity and retention in the human body should be minimal. NO has been a focus in air pollution research and thus provides a basis for work by clinicians evaluating NO inhalation in humans. In rats with chronic hypoxia- and MCT-induced PH, iNO results in a decrease in PAP with no changes in systemic arterial pressure [43–45] (**Figure 4**).

3.2 Clinical effects of iNO

iNO dilates the pulmonary vasculature by NO combining with guanylate cyclase Fe in pulmonary vascular smooth muscle cells. Most NO diffuses into the blood at alveoli, where it reacts with the Fe of oxygenated hemoglobin (oxy-Hb, O_2Hb, $O_2Hb(Fe^{2+})$) in red blood cells and is converted to NO_3^-. When NO reacts with oxy-Hb Fe^{2+} or combines with the Fe^{2+} of deoxygenated Hb (deoxy-Hb, deoxy-Hb(Fe^{2+})) in red blood cells, iNO does not have direct vasodilatory effects because it reacts or combines with Hb Fe^{2+} and becomes unable to combine with guanylate cyclase Fe of smooth muscle cells in systemic arteries. Thus, iNO is a selective pulmonary vasodilator, causing decreased PAP with no changes in systemic pressure (**Figure 4**).

The clinical use of NO inhalation is aimed at inducing selective pulmonary arterial dilation and treating PH and right ventricular failure. NO inhalation is also used to test pulmonary vascular reactivity in catheterization labs, which will be discussed in the last section of this chapter. Improved arterial oxygenation is also expected in patients with high intrapulmonary shunting [46]. Thus, the main target of iNO is lung and pulmonary circulation. In addition, iNO effects on remote organs, such as the kidney [47], liver [48, 49], heart [50], and muscle [51], have been investigated, with iNO shown to ameliorate inflammation and ischemia-reperfusion injury.

3.3 Substances that react with NO

NO reacts or combines with transition metal ions, such as thiols (–SH, –SS–, and HS–). Many enzymes and substances involved in regulating cell function include in their structure Fe, a transitional metal, thus suggesting its importance as an NO target. Because hemoglobin (Hb) and guanylate cyclase contain heme, which includes Fe in its structure, NO reacts or combines with Hb and guanylate cyclase. NO also combines with enzymes containing Fe–S in their structure, and combined NO (nitration) and Fe–S can prevent enzymatic activity. High concentrations of NO induce cell damage, which presumably result from this enzymatic dysfunction. Thus, NO is a double-edged sword. Although an appropriate amount is important for regulating cell function, an excess dysregulates cell function and causes damage.

RS–NO is a complex of SH– and NO. Nitrosothiol is a thionitrite including *S*-nitroso-albumin, where NO combines with cysteine, a component of albumin. –SH is a component of amino acids, peptides, and proteins. NO binds to –SH, forming *S*-nitrosothiol, 96% of which is *S*-nitrosoprotein. About 82% of *S*-nitrosoprotein is serum *S*-nitrosoalbumin. Thus, endogenous NO circulates in the form of *S*-nitrosoalbumin [52].

NO targets are transition metal ions, oxygen, nucleophilic centers (thiols, amides, carboxyls, and hydroxyls), and free radicals. iNO targets are (**Figure 5**) also transition metal ions, namely, in the guanylate cyclase Fe, Hb Fe, iron-sulfur (Fe-S) center. Other targets include oxygen (gaseous oxygen in the airway and alveoli), dissolved oxygen in the tissue and body fluids, the nucleophilic center of organic compounds (–S–S– and –SH), and free radicals (reactive oxygen species produced in leucocytes and macrophages). Among these substances that react with iNO, The main target of an approved medical iNO gas is guanylate cyclase Fe in pulmonary vascular smooth muscle cells.

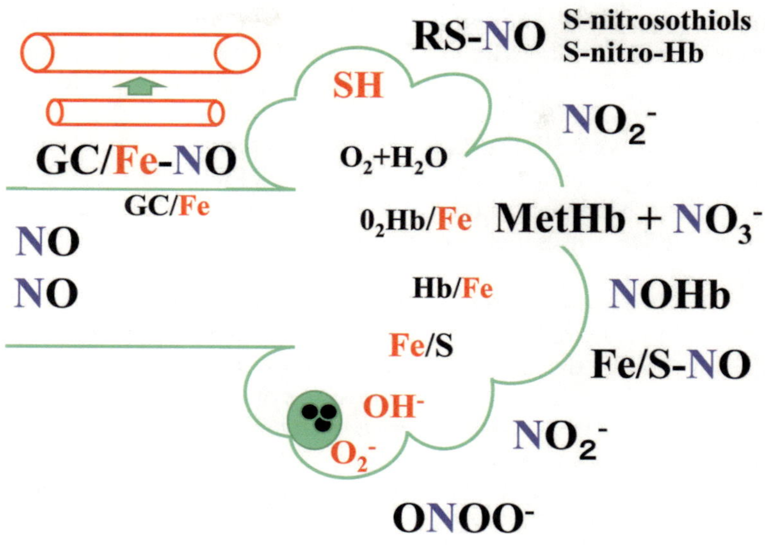

Figure 5.
The substances in the lung to react or combine with iNO. NO reacts with Fe in the guanylate cyclase (GC) and induces cyclic GMP and subsequent pulmonary vascular relaxation; NO reacts with Fe^{2+} in O_2Hb and forms MetHb and NO_3^-; NO and Fe^{2+} in deoxy-Hb combine in NOHb; NO and O_2 combine in NO_2; NO and O_2- combine in ONOO-; NO and OH- combine in NO_2-; NO and thiol (sulfhydryl group, -SH group), amine, and iron-sulfur (Fe-S) center combine in nitrosothiol, nitrosamine, and Fe-S NO, respectively. RSH is a compound including SH group. S-nitro-Hb is combination of NO and SH in the cysteine in the Hb beta subunit. Reactive oxygen species (O_2^-, OH^-) are produced in leucocytes and macrophages.

3.4 Metabolic fate of iNO

3.4.1 Oxidation and reduction of NO

Oxidation refers to electron loss and reduction to electron gain. Reducing agents release electrons, whereas oxidizing agents receive electrons. The term redox is a combination of "reduction and oxidation reaction." Reduction and oxygenation occur simultaneously so that when a reducing agent is oxidized, an oxidizing agent is also reduced. Nitrogen monoxides involve an array of species: NO^+ (nitrosonium), NO·, and NO^- (nitroxyl anion) [53]. Among these, NO· has a single electron, and its removal forms NO^+, whereas its addition yields NO^-. NO· is electrically neutral, which contributes to its free diffusibility in aqueous medium and across cell membranes.

The main NO· targets are oxygen and transition metal ions. The various redox forms of oxygen, such as superoxide (O_2^-) and (di)oxygen, are candidates in both the gas phase and aqueous solution. Metalloproteins, such as heme-containing protein and non-heme-containing protein, and iron-sulfur clusters also react with NO·.

iNO reacts with oxy-Hb. NO oxidizes oxy-Hb to form MetHb. In other words, MetHb is oxidized oxy-Hb. Oxidized iron (MetHb) species do not catch NO, and iNO during cardiopulmonary bypass (CPB) decreases acute kidney injury [47]:

$$\underset{[\text{oxy-Hb}]}{O_2Hb(Fe^{2+})} + NO \rightarrow \underset{[\text{MetHb}]}{Hb(Fe^{3+})} + NO_3^- \quad (1)$$

3.4.2 NO and Hb

NO reacts or combines with Hb in three ways: (1) NO combines with in the heme Fe to form NOHb (nitrosyl Hb), a metal nitrosyl species; (2) NO^- combines with amines in Hb to form *S*-nitroso-Hb, a nitrosamine, where NO combines with cysteine in the beta subunit of Hb; and (3) O^- or NO^+ combines with the sulfhydryl center (-SH) in Hb. NO reacts with oxy-Hb and combines with deoxy-Hb. If NO reacts with oxy-Hb (Fe^{2+}), MetHb (Fe^{3+}) and nitrate (NO_3^-) are formed. If NO combines with deoxy-Hb (Fe^{2+}), NOHb (Fe^{2+}) is formed, after which NOHb (Fe^{2+}) reacts with O_2 to form MetHb (Fe^{3+}) and NO_3^-. MetHb (Fe^{3+}) is reduced to deoxy-Hb (Fe^{2+}) by MetHb reductase. The depletion of MetHb reductase or high production of MetHb causes methemoglobinemia. NO_3^- is excreted in the urine (**Figures 6**, 7).

In an in vitro experiment, Wennmalm [54] incubated NO with arterial and venous blood and measured MetHb, NOHb, NO_3^-, and NO_2^-. The reaction of NO with O_2Hb was rapid in the arterial blood (oxygen saturation 94–99%). NOHb was low in arterial blood and high in venous blood (oxygen saturation 36–86%). These results suggest that O_2Hb (oxy-Hb) gives O_2 to NO making NO_3^-. In contrast, deoxy-Hb directly combines with NO making NOHb in the absence of oxy-Hb (i.e., in the presence of deoxy-Hb). The NO and oxy-Hb reaction is completed in 100 ms [55].

3.4.3 NOHb

Nakajima and Oda have shown that the NOHb concentration is 0.13% in the blood during 20 min of 10 ppm NO inhalation [56]. This low concentration suggests the rapid turnover of NOHb, in which NOHb is presumably an intermediate in the conversion from NO to NO_2^- and NO_3^-.

3.4.4 Conversion of NO_2^- to NO_3^-

When 5 mM NO_2^- is added to human blood, NO_3^- changes are detected within 10 min [57]. The intravenous injection of sodium nitrite to rabbits results in the rapid disappearance of NO_2^-. After the intratracheal injection of $^{13}NO_2^-$, 70% of $^{13}NO_2^-$ changed to $^{13}NO_3^-$, and 26% remained as $^{13}NO_2^-$ [58]. These observations suggest that NO_2^- is converted to NO_3^- in red blood cells [59]. NO_2^- and NO_3^- are stable and unchanged in plasma without red blood cells.

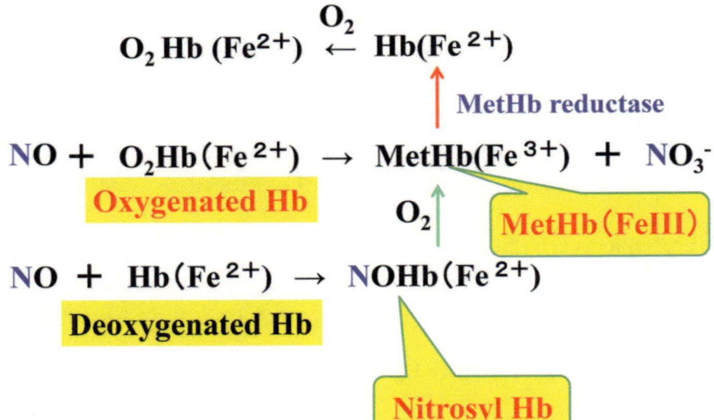

Figure 6.
Reaction of NO with Hb. NO reacts with Fe^{2+} of oxy-Hb making MetHb and NO_3^- and combines with deoxy-Hb making NOHb. NOHb reacts with O_2 making MetHb and NO_3^-.

3.4.5 No retention of iNO in the body

A clear understanding of the fate of iNO is critical for its clinical use in humans. As previously mentioned, the metabolic fate of iNO was examined in the early 1980s, and it was found that retention of iNO in the body was lacking (**Figure 7**).

An inhalation study of ^{15}NO in rats investigated the metabolism of iNO. In the carcasses, 1.6% of total inhaled ^{15}N was detected, similar to the level of natural ^{15}N. This result suggests that iNO largely does not remain in the body. About 55% of total inhaled ^{15}N was recovered in urine [41, 42], including 45% as nitrates and 10% as urea (**Figure 7**). About 10% of total inhaled ^{15}N was recovered as undetermined nitrogen compounds in feces. The remaining 35% was not recovered but is assumed to be N_2 produced from the reduction of $NO_3^- \rightarrow NO_2^- \rightarrow N_2$ by stomach flora.

3.5 iNO effects in remote organs

During cardiopulmonary bypass(CPB), hemolysis causes an increase in Hb plasma concentration due to the destruction of red blood cells. Hb includes oxy-Hb and deoxy-Hb. Oxy-Hb causes vasoconstriction, which is partly due to the depletion of NO available to induce vascular smooth muscle relaxation. NO is produced in and released from endothelial cells, some of which reaches adjacent vascular smooth muscles, causing vasorelaxation, and some of which diffuses into the plasma. If the amount of oxy-Hb in the plasma increases, the binding of NO to oxy-Hb increases, resulting in less NO reaching adjacent smooth muscle cells. Thus, the presence of large amounts of oxy-Hb might decrease NO availability in vascular smooth muscle cells (**Figure 8**).

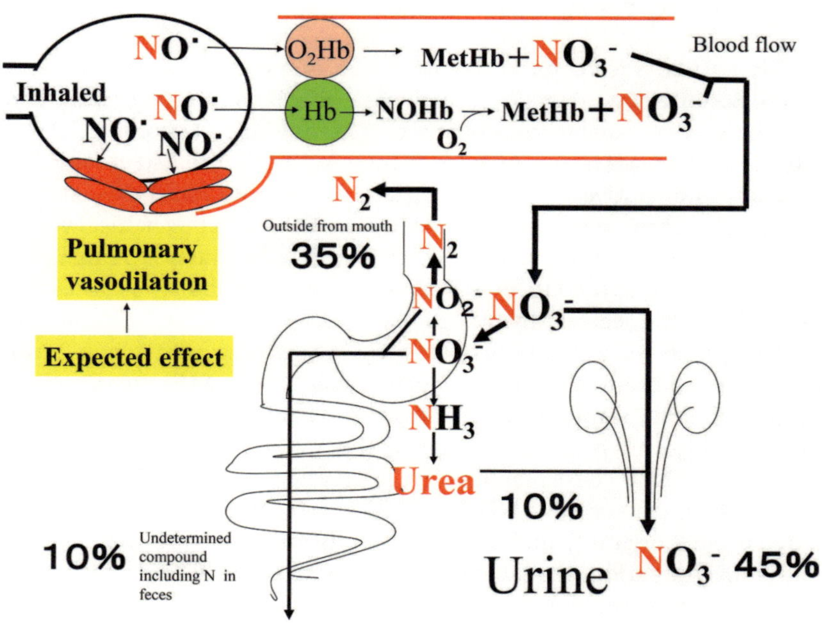

Figure 7.
Metabolic fate of NO. Almost all inhaled NO is converted to NO_3^-. Forty-five percent of NO_3^- is excreted in urine; 10% is changed to nitrogen compound except NO_3^- and NO_2^- and excreted in feces; 10% is changed to urea through the digestive tract and liver and excreted in urine. The rest will be changed to N_2 in the stomach and discharged outside of the body.

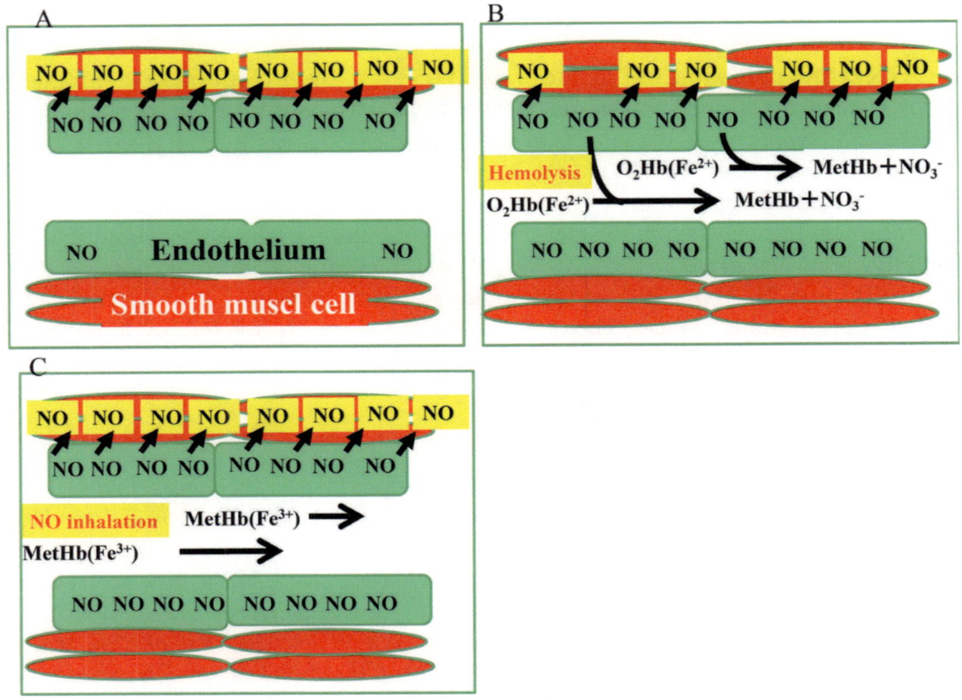

Figure 8.
NO formed in the endothelium is scavenged by plasma oxygenated Hb. (A) NO produced in the endothelium diffuses into adjacent smooth muscle cells and binds with guanylate cyclase. (B) If O_2Hb (Fe^{2+}) increases, NO produced in the endothelium diffuses into the blood and scavenged by O_2Hb (Fe^{2+}), decreasing the amount of NO diffused into smooth muscle cells. (C) If O_2 Hb(Fe^{2+}) is converted to MetHb(Fe^{3+}) by NO inhalation, NO is not scavenged, which recovers the amount of NO diffusing to the smooth muscle side.

Acute kidney injury is a common complication after cardiac surgery with prolonged CPB. Because oxygen tension is high during CPB, plasma oxy-Hb exhibits substantial hemolysis, causing vasoconstriction in the kidney. Recently, NO was demonstrated to decrease the occurrence of acute kidney injury and chronic kidney disease 1 year postoperatively [47]. NO inhalation at 80 ppm was started at the onset of CPB via a CPB machine and was continued after CPB via a mechanical ventilator for 24 h or less if patients were ready to be extubated early. Under NO inhalation, oxy-Hb was converted to MetHb, which recovered NO availability to vascular smooth muscle cells due to the decrease in oxy-Hb. Thus, exogenous NO inhalation might increase endogenous NO availability to counteract renal vasoconstriction during CPB.

In brief, the reduction of nitrite and nitrate produces NO (**Figure 9**). Nitrite is reduced by deoxy-Hb, respiratory chain enzymes, xanthine oxidoreductase, deoxygenated myoglobin, and protons, facilitating the transfer of protons to NO_2^- and thereby producing NO. These reactions are intensified under acidic and hypoxic states. After iNO is converted to NO_2^- and NO_3^-, NO can be recycled from nitrite and used to protect organs from ischemia-reperfusion injury [48, 51].

In liver transplantation, the inhalation of 80 ppm NO until reperfusion ameliorates apoptosis, attenuated increases of liver enzymes, and enhanced the recovery of coagulation factors [48].

In orthopedic knee surgery, NO inhalation prevented increases in the adhesion molecule expression on granulocytes, plasma selectin levels, and NF-κB expression in quadriceps muscles [51]. NO inhalation was started before the tourniquet application and was continued during reperfusion until the completion of surgery.

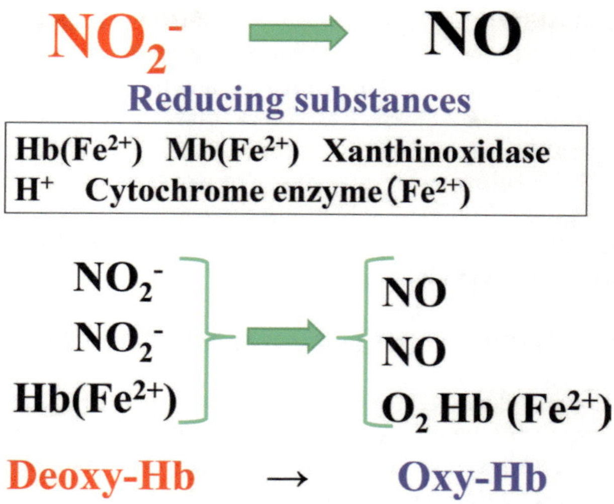

Figure 9.
Recycle of NO from nitrite and nitrate. Nitrate is reduced to nitrite, and subsequent reduction of nitrite forms NO. Reducing substances are deoxy-Hb [Hb(Fe^{2+})], myoglobin [Mb(Fe^{2+})], xanthine oxidase, hydrogen ion (H^+), and cytochrome enzymes (Fe^{2+}). NO_2^- is reduced by deoxy-Hb(Fe^{2+}) to NO showing stoichiometric relation. Please count the number of O before and after the reduction responses.

4. Update on the clinical application of iNO in pediatric patients

After NO was identified as an endothelial cell-derived relaxation factor and following preclinical studies, iNO therapy has been studied extensively in multicenter randomized trials as well as in early pilot studies of infants with severe hypoxemia associated with PH or infants with congenital diaphragmatic hernia (CDH) [60]. These studies have demonstrated improved oxygenation and reduction in the need for extracorporeal membrane oxygenation (ECMO) therapy, leading to the approval of iNO therapy by the Food and Drug Administration for use in patients at >34-week gestation with hypoxemic respiratory failure and persistent PH of the newborn (PPHN). Over the last two decades, the discussion of its application has been extended to premature infants and acute pulmonary vascular response testing to assess indications for specific pulmonary vasodilator therapy for patients with PAH or operability for children with congenital heart disease.

4.1 Role of inhaled NO in the prevention of bronchopulmonary dysplasia in premature newborns

Bronchopulmonary dysplasia (BPD), which is characterized by impaired pulmonary development resulting from insults affecting the immature lung, including inflammation, hyperoxia, and mechanical ventilation, is associated with high mortality and adverse long-term neurological and respiratory outcomes in infants born very preterm. Although the effectiveness of iNO for the treatment of PPHN is largely due to its function as a selective pulmonary vasodilator, laboratory observations also suggest other important biological effects of NO, such as roles in decreasing lung inflammation (e.g., lung vascular protein leak; pulmonary neutrophil accumulation) [61], reducing oxidant stress [62], decreasing pulmonary vascular cell proliferation [63], and enhancing alveolarization and lung growth [64–66]. These observations have led to investigations into the use of iNO to prevent the development of BPD in premature newborns. In an initial randomized, placebo-controlled study in a single center, 7 days of iNO prevented chronic lung disease in premature infants [67]. Despite promising findings in some subsequent studies showing a reduction in BPD in premature newborns [68, 69], later trials did not confirm the beneficial effects [70]. Meta-analyses of these studies have not found evidence for a net improvement in either chronic lung disease or developmental sequelae [71], leading to the conclusion by the National Institutes of Health Consensus Development Conference [72] and the American Academy of Pediatrics Committee on the Fetus and Newborn [73] that the use of iNO to prevent BPD is not supported by available evidence [74].

4.2 iNO for the treatment of severe pulmonary hypertension in preterm infants

In addition to the use of iNO for BPD prevention, its use in preterm infants for acute management of severe hypoxemic respiratory failure has been discussed. Several case series have described responses to iNO in premature newborns with PPHN associated with prolonged oligohydramnios and pulmonary hypoplasia. Chock et al. evaluated a subset of 12 premature newborns enrolled in the Preemie Inhaled Nitric Oxide Trial with pulmonary hypoplasia after preterm premature rupture of membranes (PPROM) [75]. Six infants were treated with iNO with a mortality rate of 33% compared with 67% mortality for six infants in the placebo control group. Shah and Kluckow described outcomes for infants with PPROM and reported that survival improved from 62 to 90% after the introduction of iNO and high-frequency oscillatory ventilation [76]. Semberova et al. reported a series of 22 premature infants with a history of PPROM, pulmonary hypoplasia, and PPHN who were treated with iNO, with a survival rate of 86% [77]. Thus, iNO therapy may have important benefits in subgroups of preterm infants with severe PH, especially in patients with oligohydramnios and lung hypoplasia. Further studies of the precise effects of iNO in premature neonates are needed.

4.3 Role of iNO in newborns with congenital diaphragmatic hernia

iNO in neonates with CDH has been evaluated in three randomized trials [78–80]. Finer and Barrington performed a Cochrane Review [81] of the use of iNO for respiratory failure in infants born at or near term. They concluded that while iNO might transiently improve oxygenation, its use is not recommended for infants with CDH because the risks of a composite of either death or ECMO are similar to or worse than those of controls [82].

Based on this evidence, iNO cannot be recommended for the routine treatment of PH in patients with CDH. However, iNO continues to be regularly used for CDH. Indeed, iNO was used at some point during preoperative stabilization in 36% (191/526) of infants with CDH from the population-based CAPSNet database. The ability of iNO to improve oxygenation and reduce the need for ECMO in non-CDH patients with PH explains its continued use in patients with CDH. The lack of a response to pulmonary vasodilators in CDH is speculated to be likely due to left atrial/pulmonary vein hypertension rather than to functional changes in the pulmonary arterial vasculature [83]. A recent study suggests that the response to pulmonary vasodilators in neonates with CDH may be limited by the severity of left ventricular (LV) dysfunction and/or hypoplasia, which impairs LV filling. Careful echocardiographic assessment, therefore, may guide treatment by identifying patients who may benefit from pulmonary vasodilators, including iNO [83].

4.3.1 Acute vasoreactivity testing to assess prognosis and indications for specific PH therapy

The prognosis of children with PAH has improved in the past decade owing to new therapeutic agents and aggressive treatment strategies [84]. In idiopathic or heritable PAH (I/H-PAH), acute vasodilator testing (AVT) is recommended to identify patients who have a good long-term prognosis when treated with a long-term calcium channel blocker (CCB), accounting for 7–37% of children with PAH. For example, a >20% decrease in PAP or pulmonary vascular resistance (PVR) to inhaled NO accurately predicts a subsequent response to oral vasodilators, such as nifedipine. To identify such patients, the Sitbon criteria for positive AVT, as defined by a decrease in mean PAP by ≥10 mmHg to a value of <40 mmHg with an increased or unchanged cardiac output, is commonly used in adult I/H-PAH [85]. The Sitbon criteria can also be used to identify children who are expected to show a sustained response to CCB therapy [86]. Based on these data, the use of the Sitbon criteria is advised for AVT in children. Because only half of adult responders have a long-term hemodynamic and clinical improvement in response to CCB therapy, close long-term follow-up is required [87].

4.3.2 Acute vasoreactivity testing to assess operability of PAH associated with congenital heart disease

AVT (**Figure 10**) is also used to assess operability in children with PAH associated with a systemic-to-pulmonary shunt [87, 88]. Although pulmonary vasodilators other than iNO, such as inhaled iloprost or other orally or intravenously administered compounds (e.g., sildenafil and treprostinil), can be used for AVT, iNO ± oxygen is recommended [87]. The hemodynamic change that defines a positive response to AVT

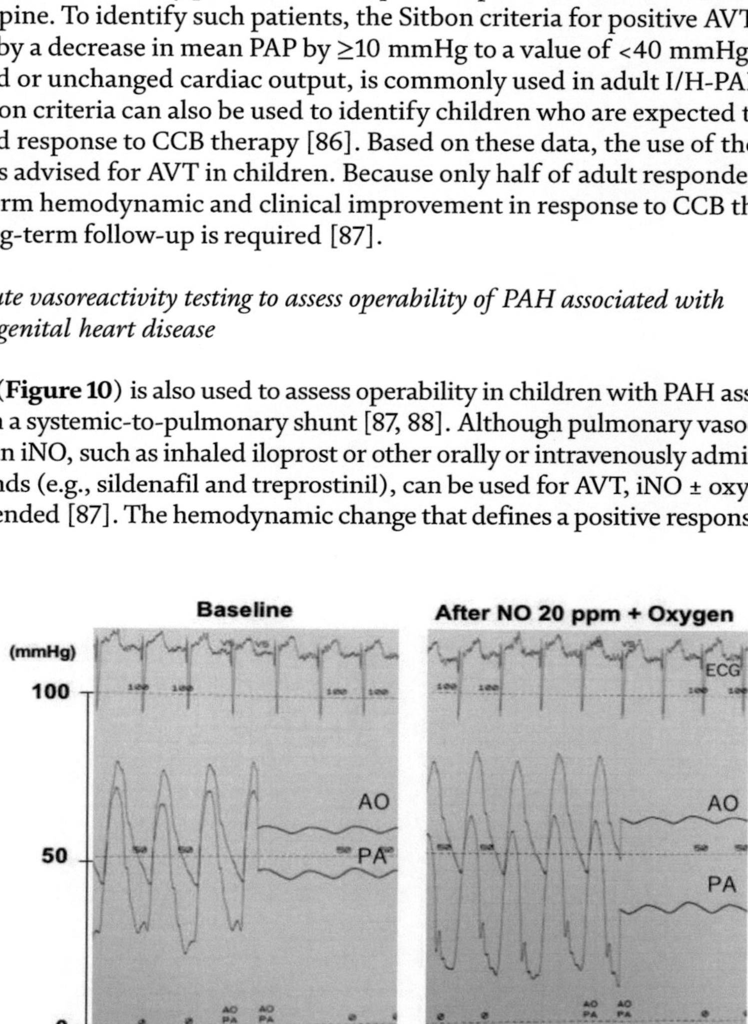

Figure 10.
*Acute vasoreactivity testingto assess operability. AVT is also used to assess operability in children with PAH associated with a systemic-to-pulmonary shunt [87, 88]. (**Figure 10**, unpublished). A 5-month-old infant with Down syndrome and an atrial septal defect was evaluated for operability by acute vasoreactivity testing using inhaled nitric oxide. Pulmonary hemodynamic parameters at baseline, including pulmonary arterial pressure (76/29/49 mmHg), pulmonary vascular resistance index (6.7 Wood units m^2), the ratio of pulmonary to systemic vascular resistance (0.45), and the ratio of pulmonary to systemic blood flow (1.74), are changed to 60/14/32 mmHg, 3.14 Wood units m^2, 0.20, and 2.2 after nitric oxide inhalation, respectively. The patient underwent surgical closure of the shunt, and no postoperative pulmonary hypertension was observed. NO, nitric oxide; AO, aorta; PA, pulmonary artery.*

in PAH associated with a shunt (a ratio of pulmonary to systemic blood flow >1.5) for children should be a >20% decrease in PVR index and a ratio of pulmonary to systemic vascular resistance with respective final values of <6 Wood units m^2 and <0.3. However, specific criteria for defining a positive AVT response that predicts the reversal of PAH and good long-term prognosis have not been described. The pediatric task force of the Sixth World Symposium on Pulmonary Hypertension agreed on a general guidance for assessing operability in CHD-PAH but emphasized that the long-term impact of defect closure in the presence of PAH with increased PVR is unknown [84].

Author details

Kazuo Maruyama[1*], Junko Maruyama[1,2] and Hirofumi Sawada[1,3]

1 Department of Anesthesiology and Critical Care Medicine, Graduate School of Medicine, Mie University, Japan

2 Faculty of Medical Engineering, Suzuka University of Medical Science, Japan

3 Department of Pediatrics, Graduate School of Medicine, Mie University, Japan

*Address all correspondence to: k-maru@clin.medic.mie-u.ac.jp

IntechOpen

© 2019 The Author(s). Licensee IntechOpen. This chapter is distributed under the terms of the Creative Commons Attribution License (http://creativecommons.org/licenses/by/3.0), which permits unrestricted use, distribution, and reproduction in any medium, provided the original work is properly cited.

References

[1] Furchgott RF, Zawadzki JV. The obligatory role of endothelial cells in the relaxation of arterial smooth muscle by acetylcholine. Nature. 1980;**288**(5789):373-376

[2] Ignarro LJ, Buga GM, Wood KS, Byrns RE, Chaudhuri G. Endothelium-derived relaxing factor produced and released from artery and vein is nitric oxide. Proceedings of the National Academy of Sciences of the United States of America. 1987;**84**(24):9265-9269

[3] Palmer RM, Ferrige AG, Moncada S. Nitric oxide release accounts for the biological activity of endothelium-derived relaxing factor. Nature. 1987;**327**(6122):524-526

[4] Katsuki S, Arnold W, Mittal C, Murad F. Stimulation of guanylate cyclase by sodium nitroprusside, nitroglycerin and nitric oxide in various tissue preparations and comparison to the effects of sodium azide and hydroxylamine. Journal of Cyclic Nucleotide Research. 1977;**3**(1):23-35

[5] Stamler JS, Loh E, Roddy MA, Currie KE, Creager MA. Nitric oxide regulates basal systemic and pulmonary vascular resistance in healthy humans. Circulation. 1994;**89**(5):2035-2040

[6] Steudel W, Ichinose F, Huang PL, Hurford WE, Jones RC, Bevan JA, et al. Pulmonary vasoconstriction and hypertension in mice with targeted disruption of the endothelial nitric oxide synthase (NOS 3) gene. Circulation Research. 1997;**81**(1):34-41

[7] Maruyama J, Maruyama K. Impaired nitric oxide-dependent responses and their recovery in hypertensive pulmonary arteries of rats. The American Journal of Physiology. 1994;**266**(6 Pt 2):H2476-H2488

[8] Maruyama J, Jiang BH, Maruyama K, Takata M, Miyasaka K. Prolonged nitric oxide inhalation during recovery from chronic hypoxia does not decrease nitric oxide-dependent relaxation in pulmonary arteries. Chest. 2004;**126**(6):1919-1925

[9] Giaid A, Saleh D. Reduced expression of endothelial nitric oxide synthase in the lungs of patients pulmonary hypertension. The New England Journal of Medicine. 1995;**333**(4):214-221

[10] Kharitonov SA, Cailes JB, Black CM, Du Bois RM, Barnes PJ. Decreased nitric oxide in the exhaled air of patients with systemic sclerosis with pulmonary hypertension. Thorax. 1997;**52**(12):1051-1055

[11] Berger RM, Geiger R, Hess J, Bogers AJ, Mooi WJ. Altered arterial expression patterns of inducible and endothelial nitric oxide synthase in pulmonary plexogenic arteriopathy caused by congenital heart disease. American Journal of Respiratory and Critical Care Medicine. 2001;**163**(6):1493-1499

[12] Zhao YY, Zhao YD, Mirza MK, Huang JH, Potula HH, Vogel SM, et al. Persistent eNOS activation secondary to caveolin-1 deficiency induces pulmonary hypertension in mice and humans through PKG nitration. The Journal of Clinical Investigation. 2009;**119**(7):2009-2018

[13] Kanno S, Wu YJ, Lee PC, Billiar TR, Ho C. Angiotensin-converting enzyme inhibitor preserves p21 and endothelial nitric oxide synthase expression in monocrotaline-induced pulmonary arterial hypertension in rats. Circulation. 2001;**104**(8):945-950

[14] Yamada Y, Maruyama J, Zhang E, Okada A, Yokochi A, Sawada H, et al. Effect of thrombomodulin on the

development of monocrotaline-induced pulmonary hypertension. Journal of Anesthesia. 2014;**28**(1):26-33

[15] Christou H, Hudalla H, Michael Z, Filatava EJ, Li J, Zhu M, et al. Impaired pulmonary arterial vasoconstriction and nitric oxide-mediated relaxation underlie severe pulmonary hypertension in the Sugen-hypoxia rat model. The Journal of Pharmacology and Experimental Therapeutics. 2018;**364**(2):258-274

[16] Oliveira SDS, Chen J, Castellon M, Mao M, Raj JU, Comhair S, et al. Injury-induced shedding of extracellular vesicles depletes endothelial cells of Cav-1 (Caveolin-1) and enables TGF-β (transforming growth factor-β)-dependent pulmonary arterial hypertension. Arteriosclerosis, Thrombosis, and Vascular Biology. 2019;**39**(6):1191-1202

[17] Mukhopadhyay S, Xu F, Sehgal PB. Aberrant cytoplasmic sequestration of eNOS in endothelial cells after monocrotaline, hypoxia, and senescence: Live-cell caveolar and cytoplasmic NO imaging. American Journal of Physiology. Heart and Circulatory Physiology. 2007;**292**(3):H1373-H1389

[18] Austin ED, Ma L, LeDuc C, Berman Rosenzweig E, Borczuk A, Phillips JA 3rd, et al. Whole exome sequencing to identify a novel gene (caveolin-1) associated with human pulmonary arterial hypertension. Circulation. Cardiovascular Genetics. 2012;**5**(3):336-343

[19] Aldred MA, Vijayakrishnan J, James V, Soubrier F, Gomez-Sanchez MA, Martensson G, et al. BMPR2 gene rearrangements account for a significant proportion of mutations in familial and idiopathic pulmonary arterial hypertension. Human Mutation. 2006;**27**(2):212-213

[20] Cogan JD, Pauciulo MW, Batchman AP, Prince MA, Robbins IM, Hedges LK, et al. High frequency of BMPR2 exonic deletions/duplications in familial pulmonary arterial hypertension. American Journal of Respiratory and Critical Care Medicine. 2006;**174**(5):590-598

[21] Thomson JR, Machado RD, Pauciulo MW, Morgan NV, Humbert M, Elliott GC, et al. Sporadic primary pulmonary hypertension is associated with germline mutations of the gene encoding BMPR-II, a receptor member of the TGF-beta family. Journal of Medical Genetics. 2000;**37**(10):741-745

[22] Hartung A,Bitton-Worms K, Rechtman MM, Wenzel V, Boergermann JH, Hassel S, et al. Different routes of bone morphogenic protein (BMP) receptor endocytosis influence BMP signaling. Molecular and Cellular Biology. 2006;**26**(20):7791-7805

[23] Wertz JW, Bauer PM. Caveolin-1 regulates BMPRII localization and signaling in vascular smooth muscle cells. Biochemical and Biophysical Research Communications. 2008;**375**(4):557-561

[24] Prewitt AR, Ghose S, Frump AL, Datta A, Austin ED, Kenworthy AK, et al. Heterozygous null bone morphogenetic protein receptor type 2 mutations promote SRC kinase-dependent caveolar trafficking defects and endothelial dysfunction in pulmonary arterial hypertension. The Journal of Biological Chemistry. 2015;**290**(2):960-971

[25] Lantuéjoul S, Sheppard MN, Corrin B, Burke MM, Nicholson AG. Pulmonary veno-occlusive disease and pulmonary capillary hemangiomatosis: A clinicopathologic study of 35 cases. The American Journal of Surgical Pathology. 2006;**30**(7):850-857

[26] Kradin R, Matsubara O, Mark EJ. Endothelial nitric oxide synthase expression in pulmonary capillary hemangiomatosis. Experimental and Molecular Pathology. 2005;**79**(3):194-197.21

[27] Eyries M, Montani D, Girerd B, Perret C, Leroy A, Lonjou C, et al. EIF2AK4 mutations cause pulmonary veno-occlusive disease, a recessive form of pulmonary hypertension. Nature Genetics. 2014;**46**(1):65-69

[28] Perros F, Günther S, Ranchoux B, Godinas L, Antigny F, Chaumais MC, et al. Mitomycin-induced pulmonary veno-occlusive disease: Evidence from human disease and animal models. Circulation. 2015;**132**(9):834-847

[29] Nossent EJ, Antigny F, Montani D, Bogaard HJ, Ghigna MR, Lambert M, et al. Pulmonary vascular remodeling patterns and expression of general control nonderepressible 2 (GCN2) in pulmonary veno-occlusive disease. The Journal of Heart and Lung Transplantation. 2018;**37**(5):647-655

[30] Francis BN, Hale A, Channon KM, Wilkins MR, Zhao L. Effects of tetrahydrobiopterin oral treatment in hypoxia-induced pulmonary hypertension in rat. Pulmonary Circulation. 2014;**4**(3):462

[31] Murata T, Kinoshita K, Hori M, Kuwahara M, Tsubone H, Karaki H, et al. Statin protects endothelial nitric oxide synthase activity in hypoxia-induced pulmonary hypertension. Arteriosclerosis, Thrombosis, and Vascular Biology. 2005;**25**(11):2335-2342

[32] Blumberg FC, Wolf K, Arzt M, Lorenz C, Riegger GA, Pfeifer M. Effects of ET-A receptor blockade on eNOS gene expression in chronic hypoxic rat lungs. Journal of Applied Physiology (Bethesda, MD: 1985). 2003;**94**(2):446-452

[33] Kim SY, Lee JH, Huh JW, Kim HJ, Park MK, Ro JY, et al. Bortezomib alleviates experimental pulmonary arterial hypertension. American Journal of Respiratory Cell and Molecular Biology. 2012;**47**(5):698-708

[34] Chalupsky K, Kračun D, Kanchev I, Bertram K, Görlach A. Folic acid promotes recycling of tetrahydrobiopterin and protects against hypoxia-induced pulmonary hypertension by recoupling endothelial nitric oxide synthase. Antioxidants & Redox Signaling. 2015;**23**(14):1076-1091

[35] Bei Y, Duong-Quy S, Hua-Huy T, Dao P, Le-Dong NN, Dinh-Xuan AT. Activation of RhoA/Rho-kinase pathway accounts for pulmonary endothelial dysfunction in patients with chronic obstructive pulmonary disease. Physiological Reports. 2013;**1**(5):e00105

[36] Yang Q, Shigemura N, Underwood MJ, Hsin M, Xue HM, Huang Y, et al. NO and EDHF pathways in pulmonary arteries and veins are impaired in COPD patients. Vascular Pharmacology. 2012;**57**(2-4):113-118

[37] Almudéver P, Milara J, De Diego A, Serrano-Mollar A, Xaubet A, Perez-Vizcaino F, et al. Role of tetrahydrobiopterin in pulmonary vascular remodelling associated with pulmonary fibrosis. Thorax. 2013;**68**(10):938-948

[38] Higenbottam T, Pepke-Zeba J, Scott J, et al. Inhaled endothelium derived-relaxing factor (EDRF) in primary pulmonary hypertension (PPH). The American Review of Respiratory Disease. 1988;**137**(Suppl):A103

[39] Pepke-Zaba J, Higenbottam TW, Dinh-Xuan AT, Stone D, Wallwork J. Inhaled nitric oxide as a cause of selective pulmonary vasodilatation in pulmonary

hypertension. Lancet. 1991;**338**(8776):1173-1174

[40] Frostell C, Fratacci MD, Wain JC, Jones R, Zapol WM. Inhaled nitric oxide. A selective pulmonary vasodilator reversing hypoxic pulmonary vasoconstriction. Circulation. 1991;**83**(6):2038-2047

[41] Yoshida K, Kasama K, Kitabatake M, Okuda M, Imai M. Metabolic fate of nitric oxide. International Archives of Occupational and Environmental Health. 1980;**46**(1):71-77

[42] Yoshida K, Kasama K, Kitabatake M, Imai M. Biotransformation of nitric oxide, nitrite and nitrate. International Archives of Occupational and Environmental Health. 1983;**52**(2):103-115

[43] Jiang BH, Maruyama J, Yokochi A, Amano H, Mitani Y, Maruyama K. Correlation of inhaled nitric-oxide induced reduction of pulmonary artery pressure and vascular changes. The European Respiratory Journal. 2002;**20**(1):52-58

[44] Maruyama J, Maruyama K, Mitani Y, Kitabatake M, Yamauchi T, Miyasaka K. Continuous low-dose NO inhalation does not prevent monocrotaline-induced pulmonary hypertension in rats. The American Journal of Physiology. 1997;**272** (1 Pt 2):H517-H524

[45] Maruyama K, Maruyama J, Muneyuki M. Nitric oxide in pulmonary circulation. Japanese Journal of Intensive Care Medicine. 1994;**18**:1049-1057

[46] Ichinose F, Roberts JD Jr, Zapol WM. Inhaled nitric oxide: A selective pulmonary vasodilator: Current uses and therapeutic potential. Circulation. 2004;**109**(25):3106-3111

[47] Lei C, Berra L, Rezoagli E, Yu B, Dong H, Yu S, et al. Nitric oxide decreases acute kidney injury and stage 3 chronic kidney disease after cardiac surgery. American Journal of Respiratory and Critical Care Medicine. 2018;**198**(10):1279-1287

[48] Lang JD Jr, Teng X, Chumley P, Crawford JH, Isbell TS, Chacko BK, et al. Inhaled NO accelerates restoration of liver function in adults following orthotopic liver transplantation. The Journal of Clinical Investigation. 2007;**117**(9):2583-2591

[49] Lang JD Jr, Smith AB, Brandon A, Bradley KM, Liu Y, Li W, et al. A randomized clinical trial testing the anti-inflammatory effects of preemptive inhaled nitric oxide in human liver transplantation. PLoS One. 2014;**9**(2):e86053

[50] Gianetti J, Del Sarto P, Bevilacqua S, Vassalle C, De Filippis R, Kacila M, et al. Supplemental nitric oxide and its effect on myocardial injury and function in patients undergoing cardiac surgery with extracorporeal circulation. The Journal of Thoracic and Cardiovascular Surgery. 2004;**127**(1):44-50

[51] Mathru M, Huda R, Solanki DR, Hays S, Lang JD. Inhaled nitric oxide attenuates reperfusion inflammatory responses in humans. Anesthesiology. 2007;**106**(2):275-282

[52] Stamler JS, Jaraki O, Osborne J, Simon DI, Keaney J, Vita J, et al. Nitric oxide circulates in mammalian plasma primarily as an S-nitroso adduct of serum albumin. Proceedings of the National Academy of Sciences of the United States of America. 1992;**89**(16):7674-7677

[53] Stamler JS, Singel DJ, Loscalzo J. Biochemistry of nitric oxide and its redox-activated forms. Science. 1992;**258**(5090):1898-1902

[54] Wennmalm A, Benthin G, Petersson AS. Dependence of the metabolism of nitric oxide (NO) in healthy human whole blood on the oxygenation of its red cell haemoglobin. British Journal of Pharmacology. 1992;**106**(3):507-508

[55] Kelm M, Schrader J. Control of coronary vascular tone by nitric oxide. Circulation Research. 1990;**66**(6):1561-1575

[56] Oda H, Kusumoto S, Nakajima T. Nitrosyl-hemoglobin formation in the blood of animals exposed to nitric oxide. Archives of Environmental Health. 1975;**30**(9):453-456

[57] Kelm M, Feelish M, Grube W, et al. Metabolism of endothelium-derived nitric oxide in human blood. In: Moncada S, editor. Biology of Nitric Oxide. Colchester: Portland Press; 1994. pp. 319-322

[58] Parks NJ, Krohn KJ, Mathis CA, Chasko JH, Geiger KR, Gregor ME, et al. Nitrogen-13-labeled nitrite and nitrate: Distribution and metabolism after intratracheal administration. Science. 1981;**212**(4490):58-60

[59] Kosaka H, Uozumi M, Tyuma I. The interaction between nitrogen oxides and hemoglobin and endothelium-derived relaxing factor. Free Radical Biology & Medicine. 1989;**7**(6):653-658

[60] Barrington KJ, Finer N, Pennaforte T, Altit G. Nitric oxide for respiratory failure in infants born at or near term. Cochrane Database of Systematic Reviews. 2017;(1):Cd000399

[61] Kinsella JP, Parker TA, Galan H, Sheridan BC, Halbower AC, Abman SH. Effects of inhaled nitric oxide on pulmonary edema and lung neutrophil accumulation in severe experimental hyaline membrane disease. Pediatric Research. 1997;**41**:457-463

[62] Nelin LD, Welty SE, Morrisey JF, Gotuaco C, Dawson CA. Nitric oxide increases the survival of rats with a high oxygen exposure. Pediatric Research. 1998;**43**:727-732

[63] Roberts JD Jr, Chiche JD, Weimann J, Steudel W, Zapol WM, Bloch KD. Nitric oxide inhalation decreases pulmonary artery remodeling in the injured lungs of rat pups. Circulation Research. 2000;**87**:140-145

[64] Lin Y-J, Markham NE, Balasubramaniam V, Tang J-R, Maxey A, Kinsella JP, et al. Inhaled nitric oxide enhances distal lung growth after exposure to hyperoxia in neonatal rats. Pediatric Research. 2005;**58**:22-29

[65] McCurnin DC, Pierce RA, Chang LY, Gibson LL, Osborne-Lawrence S, Yoder BA, et al. Inhaled NO improves early pulmonary function and modifies lung growth and elastin deposition in a baboon model of neonatal chronic lung disease. American Journal of Physiology-Lung Cellular and Molecular Physiology. 2005;**288**:L450-L459

[66] Tang J-R, Seedorf G, Balasubramaniam V, Maxey A, Markham N, Abman SH. Early inhaled nitric oxide treatment decreases apoptosis of endothelial cells in neonatal rat lungs after vascular endothelial growth factor inhibition. American Journal of Physiology-Lung Cellular and Molecular Physiology. 2007;**293**:L1271-L1280

[67] Schreiber MD, Gin-Mestan K, Marks JD, Huo D, Lee G, Srisuparp P. Inhaled nitric oxide in premature infants with the respiratory distress syndrome. The New England Journal of Medicine. 2003;**349**:2099-2107

[68] Kinsella JP, Cutter GR, Walsh WF, Gerstmann DR, Bose CL, Hart C, et al. Early inhaled nitric oxide therapy in premature newborns with respiratory failure. New England Journal of Medicine. 2006;**355**:354-364

[69] Ballard RA, Truog WE, Cnaan A, Martin RJ, Ballard PL, Merrill JD, et al. Inhaled nitric oxide in preterm infants undergoing mechanical ventilation. The New England Journal of Medicine. 2006;**355**:343-353

[70] MercierJC,HummlerH,DurrmeyerX, Sanchez-Luna M, Carnielli V, Field D, et al. Inhaled nitric oxide for prevention of bronchopulmonary dysplasia in premature babies (EUNO): A randomised controlled trial. Lancet. 2010;**376**:346-354

[71] Askie LM, Ballard RA, Cutter G, Dani C, Elbourne D, Field D, et al. Inhaled nitric oxide in preterm infants: A systematic review and individual patient data meta-analysis. BMC Pediatrics. 2010;**10**:15

[72] Cole FS, Alleyne C, Barks JD, Boyle RJ, Carroll JL, Dokken D, et al. NIH consensus development conference statement: Inhaled nitric-oxide therapy for premature infants. Pediatrics. 2011;**127**:363-369

[73] Kumar P. Use of inhaled nitric oxide in preterm infants. Pediatrics. 2014;**133**:164-170

[74] Kinsella JP, Steinhorn RH, Krishnan US, Feinstein JA, Adatia I, Austin ED, et al. Recommendations for the use of inhaled nitric oxide therapy in premature Newborns with severe pulmonary hypertension. The Journal of Pediatrics. 2016;**170**:312-314

[75] Chock VY, Van Meurs KP, Hintz SR, Ehrenkranz RA, Lemons JA, Kendrick DE, et al. Inhaled nitric oxide for preterm premature rupture of membranes, oligohydramnios, and pulmonary hypoplasia. American Journal of Perinatology. 2009;**26**:317-322

[76] Shah DM, Kluckow M. Early functional echocardiogram and inhaled nitric oxide: Usefulness in managing neonates born following extreme preterm premature rupture of membranes (PPROM). Journal of Paediatrics and Child Health. 2011;**47**:340-345

[77] Semberova J, O'Donnell SM, Franta J, Miletin J. Inhaled nitric oxide in preterm infants with prolonged preterm rupture of the membranes: A case series. Journal of Perinatology. 2015;**35**:304

[78] Kinsella JP, Truog WE, Walsh WF, Goldberg RN, Bancalari E, Mayock DE, et al. Randomized, multicenter trial of inhaled nitric oxide and high-frequency oscillatory ventilation in severe, persistent pulmonary hypertension of the newborn. The Journal of Pediatrics. 1997;**131**:55-62

[79] Group NINOS. Inhaled nitric oxide and hypoxic respiratory failure in infants with congenital diaphragmatic hernia. Pediatrics. 1997;**99**:838-845

[80] Clark RH, Kueser TJ, Walker MW, Southgate WM, Huckaby JL, Perez JA, et al. Low-dose nitric oxide therapy for persistent pulmonary hypertension of the newborn. The New England Journal of Medicine. 2000;**342**:469-474

[81] Finer NN, Barrington KJ. Nitric oxide for respiratory failure in infants born at or near term. Cochrane Database of Systematic Reviews. 2006:Cd000399

[82] Puligandla PS, Grabowski J, Austin M, Hedrick H, Renaud E, Arnold M, et al. Management of congenital diaphragmatic hernia: A systematic review from the APSA outcomes and evidence based practice committee. Journal of Pediatric Surgery. 2015;**50**:1958-1970

[83] Kinsella JP, Ivy DD, Abman SH. Pulmonary vasodilator therapy in congenital diaphragmatic hernia: Acute, late, and chronic pulmonary hypertension. Seminars in Perinatology. 2005;**29**:123-128

[84] Rosenzweig EB, Abman SH, Adatia I, Beghetti M, Bonnet D, Haworth S, et al. Paediatric pulmonary arterial hypertension: Updates on definition, classification, diagnostics and management. The European Respiratory Journal. 2019;**53**:1801916

[85] Sitbon O, Humbert M, Jais X, Ioos V, Hamid AM, Provencher S, et al. Long-term response to calcium channel blockers in idiopathic pulmonary arterial hypertension. Circulation. 2005;**111**:3105-3111

[86] Douwes JM, Humpl T, Bonnet D, Beghetti M, Ivy DD, Berger RM. Acute vasodilator response in pediatric pulmonary arterial hypertension: Current clinical practice from the TOPP registry. Journal of the American College of Cardiology. 2016;**67**:1312-1323

[87] Apitz C, Hansmann G, Schranz D. Hemodynamic assessment and acute pulmonary vasoreactivity testing in the evaluation of children with pulmonary vascular disease. Expert consensus statement on the diagnosis and treatment of paediatric pulmonary hypertension. The European paediatric pulmonary vascular disease network, endorsed by ISHLT and DGPK. Heart. 2016;**102**(Suppl. 2):ii23-ii29

[88] Balzer DT, Kort HW, Day RW, Corneli HM, Kovalchin JP, Cannon BC, et al. Inhaled nitric oxide as a preoperative test (INOP Test I): The INOP Test Study Group. Circulation. 2002;**106**:I76-I81

Section 2

Clinical Aspects of Microcirculation

Chapter 5

Endothelial Dysfunction in Cardiovascular Diseases

Indranil Biswas and Gausal A. Khan

Abstract

Endothelium is the inner most cell layer of blood vessels. Endothelial cells make special barrier that separate blood from extravascular tissues. Intact endothelium regulates vascular tone and permeability and maintains non-inflammatory, anti-thrombotic surface. Through its ability to express pro-coagulants, anticoagulants, vasoconstrictors, vasodilators, cell adhesion molecules, and cytokines, the endothelium has emerged as one of the pivotal regulators of vascular homeostasis. Under physiological conditions, endothelial cell sustains a vasodilatory, anticoagulant, and fibrinolytic state in which coagulation, platelet adhesion, as well as leukocyte activation and inflammation are suppressed. In contrast, during endothelial disturbances, a prothrombotic and pro-inflammatory state of vasoconstriction gets support from the endothelial surface. Release of platelet-activating factor (PAF) and endothelin-1 promotes vasoconstriction, whereas production of von Willebrand factor (vWF), tissue factor (TF), and plasminogen activator inhibitor (PAI)-1 shifts the haemostatic balance towards a procoagulant state. Several factors like infection, hyperglycaemia, hyperlipidaemia, malignancy, oxidative stress, and aging can interfere in endothelial function. It is believed that most of the cardiovascular diseases originate from endothelial dysfunction. Endothelial dysfunction has been shown to be involved in atherosclerosis, thrombosis, hypertension, diabetes, and other cardiovascular diseases. In this review we will specifically highlight the role of endothelial dysfunction in development of cardiovascular diseases.

Keywords: endothelial dysfunction, atherosclerosis, hypertension, heart failure, stroke

1. Introduction

Vascular endothelium is considered as a largest endocrinal organ in the body, which has been shown to have a role in homeostasis in the body by exerting various functions [1]. It is made up of simple squamous epithelial cells that line blood vessels, lymphatic vessels, and the heart. The vascular endothelium has a total weight of about 1.5 kg. The endothelium has been recognized as a smart barrier and a key regulator of blood flow in micro- and macro-vascular circulation [2]. Endothelial function is very important, as it interacts with nearly every system in the body and selectively supplies nutrients and growth factors to every organ. On the other hand, endothelium also receives active metabolites and delivers them back to the circulation. Previously, it was believed that, endothelium is an inactive barrier between blood and extravascular tissues. However, recent research has shown that the vascular endothelium is an active paracrine, endocrine, and autocrine organ that is indispensable for the regulation of vascular tone and the maintenance of vascular homeostasis.

2. Physiological function of endothelium

When immediate surrounding tissues are at basal conditions, the endothelium functions to maintain the vessel homeostasis, which favors vessel dilatation over vasoconstriction [3]. The endothelium being a dynamic reactive tissue, responds to various intrinsic physical stimuli, that is, shear stress, temperature, and transmural pressure and external stimuli, that is, temperature, mental stress, neurohumoral responses, immune response, and medications [2, 4].

Under normal conditions, endothelial cell maintains basal perfusion, which is determined by cardiac output, systemic, and local vascular resistance. Endothelial metabolism, which is a key regulator of perfusion, is impaired during several disease states like infection, injury, aging, and inflammation [5]. Local blood flow is the result of vascular relaxation and contraction that is balanced by endothelium-derived vaso-dilatative and vaso-constrictive factors. Among these factors, one signal molecule stands out as hub and target of many pathways and mechanisms that is nitric oxide (NO) [6]. It is important to understand the biochemical foundations of NO for endothelial function. NO, a potent vasodilator, is released form the endothelium due to shear stress. This NO is released by endothelial nitric oxide synthase (eNOS) by utilizing L-arginine as a substrate, which leads to the production of intracellular cyclic GMP [7]. However, in an event when the NO-dependent vasodilator mechanism is compromised, then the cytochrome-derived factors, natriuretic peptide [8], and prostacyclin [9]-dependent vasodilator mechanism came in action.

During disease state, there is impaired endothelial function, and this results in the balance shift towards prevailing constrictive factors and/or down-regulation of vaso-dilatative factors. An important counterweight in the vascular balance is cyclooxygenase (COX). This mostly induces COX-1 which is endogenous and may involve COX-2 if it is induced. The COXs have a key role in generating vasoconstrictive factors.

The COX enzymes transform arachidonic acid into endoperoxides and further into thromboxane A_2 (TXA_2) [10], prostaglandins, and prostacyclin [11]. Local presence of thrombin evokes inducible NO release. Platelet release of serotonin and ADP in turn increases NO synthesis and release in healthy endothelium to induce dilatation [12]. When vasodilatory function of endothelium is impaired, then the thrombus formation is mechanically promoted by vasoconstriction via TXA_2 and by the direct effect of serotonin on smooth muscle cells [13].

3. Endothelial dysfunction

Traditionally, endothelial dysfunction has been associated to pathological conditions that have altered anticoagulant function, impaired anti-inflammatory properties of the endothelium, impaired modulation of vascular growth, and dysregulation of vascular remodeling. For instance, a plethora of studies have confirmed the impairment of endothelium-dependent vasorelaxation caused by a loss of NO bioactivity/availability in the vessel wall [4]. The loss of NO bioavailability is the salient feature of a dysfunctional endothelium, which in turn is the sentinel of systemic or focal vascular disease. A number of previous studies have showed that most of the cardiovascular diseases were initiated from endothelial dysfunction. The decline in NO bioavailability may be caused by decreased expression of the endothelial cell eNOS [14], a lack of substrate or cofactors for eNOS [15], the presence of inhibitor of NOS [16], and alterations of cellular signaling, and finally, accelerated NO degradation by ROS [17]. Another aspect of endothelial dysfunction is impaired endothelial barrier function. Depending on the mode of pathophysiological change, this loss in barrier function may be localized or systemic. Localized loss of the selective barrier function (manifested as edema), coupled with the emigration of leukocytes, has been recognized as cardinal signs of inflammation [18]. From an immunological context, the body's primary reaction to tissue injury or infection is the leukocyte interacting with endothelium. However, from the perspective of hemostasis and thrombosis, endothelial dysfunction is characterized by activation of pro-inflammatory and pro-coagulant molecules, as well as the suppression of anti-inflammatory and anti-coagulant molecules. The intact and normal functioning endothelial lining provides a stable reservoir for blood as its luminal surface does not activate the coagulation cascade or promote leukocyte-platelet adhesion, and it also exhibits anticoagulant and fibrinolytic properties [19]. Systemic endothelial dysfunction may lead to widespread inflammation, vascular leakage, thrombocytopenia, and disseminated intravascular coagulation (DIC). On the other hand, localized endothelial dysfunction and leukocyte adhesion may lead to venous thrombosis. Other than altered endothelial barrier function, localized endothelial dysfunction also leads to tissue factor induction and increased von Willebrand factor (vWF) release that shifts the homeostatic balance towards the pro-coagulant-pro-inflammatory phenotype [20]. Intact endothelium release pro-fibrinolytic molecules like tissue plasminogen activator (TPA) [21]. Endothelial dysfunction suppressed TPA release, thereby impairing fibrinolytic function of endothelium [22]. In contrast to venous endothelial cells and microvascular endothelial cells, arterial endothelial cells are surrounded by a vascular smooth muscle layer and adventitial layer. Arterial endothelial cells physiologically experience high sheer stress and synthesize ample amount of NO that facilitate vascular relaxation. In the context of atherogenesis, endothelial cell dysfunction is mainly characterized by a loss of anatomical integrity of the intima, as described by the seminal "*Response-to-Injury Hypothesis*". Endothelial cell injury and subsequent sub-endothelial matrix exposure lead to platelet adhesion and activation mediated through sub-endothelial collagen layer [23]. The initiating event in the atherogenic process is some form of overt injury to the intimal endothelial lining that is induced by noxious substances (e.g., oxidized cholesterol, cigarette smoke, hyperlipidemia, hypercholesterolemia, hyperglycemia, etc.) or altered hemodynamic sheer stress (e.g., abnormal blood flow caused by hypertension) [24]. In particular, local endothelial mechanical tearing was seen as the inciting stimulus for platelet adhesion, activation, and the localized release of platelet-derived growth factors. This would then elicit the migration, proliferation, and phenotypic modulation of medial smooth muscle cells and thus generate a fibromuscular plaque [25]. It is of great interest to establish the sequential event that leads to atherogenesis from endothelial injury. But, the direct link between endothelial injury and the genesis of atherosclerotic lesion is still unclear. However, the detailed morphologic examination in diet-induced fatty streak lesions in animal models failed to demonstrate unconcealed intimal injury or platelet adhesion. In this context, it is highly relevant that several molecules including high mobility group protein (HMGB-1) [26] and heat shock proteins (HSPs) [27] are released from injured endothelial cells and facilitate monocyte adhesion, a crucial step for plaque formation.

4. Endothelial dysfunction in atherosclerosis

Endothelial dysfunction of lesion-prone areas of the arterial vasculature leads to atherosclerotic plaque formation [28]. Sequential deterioration of arterial vasculature along with increased sheer stress contributes to lesion formation. Endothelial dysfunction is one of the early events that are responsible for the deterioration of arterial vasculature [29]. Recent insight into the cellular mechanisms involved in atherogenesis shows that deleterious modifications of endothelial physiology or metabolism is the initial event of vascular remodeling that represents a crucial step in the development of atherosclerosis and are also involved in the development of plaque and the occurrence of atherosclerosis [2]. The sequential event including focal permeation, trapping, and physicochemical modification of circulating lipoprotein particles in the sub-endothelial space constructs an inflammatory lesion [30]. This initiates a coordinated cellular signaling, followed by complex pathogenic sequence and endothelial activation. Activated endothelial cells express several cell adhesion molecules, which facilitate selective recruitment of circulating monocytes from the blood, and invade the tunica intima, where they differentiate into macrophages. These macrophages also abnormally take up modified lipoproteins to become foam cells (the hallmark of early fatty streak lesions) [31, 32]. The activated endothelium and macrophages release multiple chemokine and growth factors which act on neighboring smooth muscle cells (or precursors cell) [33] to induce their proliferation and synthesis of extracellular matrix components within the intimal compartment, thus generating a fibromuscular plaque34. This progressive structural remodeling of developing lesions results in the formation of a fibrous cap, overlying a lipid-rich necrotic core that consists of oxidized lipoproteins, cholesterol crystals, and cellular debris. This is also accompanied by varying degrees of matrix remodeling and calcification [34, 35]. The lateral edges of these complicated plaques also contain a rich population of inflammatory cells, that is, activated macrophages, T-lymphocyte, and dendritic cells [36], which secrete several cytokines and chemokines that further activate endothelial pro-inflammatory phenotype and contribute to structural instability of the plaque through release of proteolytic enzymes (matrix metalloproteases) that lead to modification of sub-endothelial matrix components [37]. Another aspect of atherogenesis is governed by lipoproteins, mainly low-density lipoproteins (LDL). This initial arterial remodeling through accumulation of lipids is known as fatty streak formation. The first changes in the arterial wall occur at the branch points of arteries, where adaptive intimal thickening occurs in response to normal hemodynamic stresses [38]. During the early stage of atherogenesis, LDL particles leave the blood and enter the arterial intima, composed of endothelial cells. Fat droplets (LDL) may also accumulate in the cytoplasm of vascular smooth muscle cells (VSMCs) [35]. LDL particles are then modified by enzymes and are oxidized into a highly reactive pro-inflammatory molecule (oxidized LDL), that is recognized by pattern recognition receptors, that is, Toll-like receptors (TLRs) present in endothelial cells as well as pro-inflammatory macrophages [39]. Oxidized LDL incites the reaction of the innate inflammatory system within the intima and contributes to vascular remodeling. Inflammation begins when activated endothelial cells (through TLRs) express cell adhesion molecules and VSMCs secrete chemokines and chemoattractants, which together draw monocytes, lymphocytes, mast cells, and neutrophils into the arterial wall [40]. Once monocytes enter into the arterial wall through the intima, they become activated into macrophages. These macrophages take up lipids as multiple small inclusions and become transferred into foam cells [41]. The degree of lipid accumulation is critical for early stage diagnosis of atherosclerosis. Atherosclerosis is believed to start when the lipid accumulation appears as confluent extracellular lipid pools and extracellular lipid cores with and decreased cellularity [42]. Endothelial cell dysfunction is also responsible for VSMC proliferation and differentiation to myofibroblast. In an intact vessel, VSMCs never come into contact with plasma proteins and are therefore devoid of growth factor present in plasma. In physiological conditions, VSMCs are always maintained in quiescent states. During early inflammation and endothelial cell activation, VSMCs receive signal from dying cells or growth factors that modify VSMCs to myofibroblast (more proliferative counterpart). Altered VSMCs (myofibroblast) also secrete proteoglycans, collagen, and elastic fibers into the sub-endothelial matrix [43]. This transformation of VSMCs further worsens the histological structure and leads to formation of thin-cap fibroatheroma formation [44]. Fibroatheroma can be of two different types depending on the content and stability of the plaque. Stability of the plaque also determines the fate of the fibroatheroma. Unstable fibroatheroma leads to thrombotic plaque formation, whereas stable fibroatheroma accumulates calcium, becomes stiff, and eventually leads to occlusion [44, 45]. Unstable or vulnerable plaques may lead to a catastrophic transition into atherosclerotic lesion—frank plaque rupture, with luminal release of the highly thrombogenic contents [46, 47]. Else, some significant clinical sequelae can be seen from superficial intimal erosions, without any indication of plaque rupture [48]. Therefore, lately an acute transition appears leading to endothelial cell apoptosis, with localized endothelial denudation and thrombus formation leading to obstruction in regional blood flow [49, 50], whereas the stable lesions, having thick fibrous cap and less lipid as well as inflammatory cell content, can gradually invade on the lumen of the vessel causing ischemic symptoms without atherothrombotic events [51, 52]. Many ruptures of thin fibrous caps are clinically silent in that they heal by forming fibrous tissue matrices of cells, collagen fibers, and extracellular space but may rupture again with thrombus formation of the necrotic core, triggering an atherothrombotic occlusion. These cyclic changes of rupture, thrombosis, and healing may recur as many as four times at a single site in the arterial wall, resulting in multiple layers of healed tissue. In all these steps, calcium deposition in the wall of the vessels forms small aggregates initially, which turns into large nodules at later stage. In later stage, these plaques may rupture and expose the nodules, and it became sites for thrombus formation [47]. Therefore, the increasing number of plaques itself might become adequate to form significant stenosis, which may cause a shattering ischemic event due to flow restriction [53]. Based on its multiple regulatory roles throughout this complex series of events, it is evident that endothelial dysfunction constitutes a well-coordinated multicellular pathogenic sequence that leads to atherosclerosis.

5. Endothelial dysfunction in hypertension

Hypertension affects significantly to worldwide cardiovascular morbidity and mortality and is considered as a diagnostic factor for cardiovascular disease. Hypertension appears to have a complex association with endothelial dysfunction, a phenotypical alteration of the vascular endothelium that precedes the development of adverse cardiovascular events. Endothelial cells along with the vascular smooth muscle cells of resistance vessels (arteries and arterioles) regulate hypertension [54] as they continuously constrict and dilate according to the rhythm of cardiac cycle. In response to the blood flow (perfusion), the quiescent healthy endothelium continuously releases potent vasodilators, which have the potential to lower vascular resistance, thereby regulating the blood pressure [55]. In normal condition, basal perfusion is determined by cardiac output, systemic, and local resistance. In an intact healthy vessel, endothelial cell always maintains a vasodilatory rather than a vasoconstrictive phenotype. Endothelial dysfunction is a condition comprising not only attenuated endothelium-dependent vasodilatation but also an augmented inflammatory endothelial activation that leads to vasoconstriction. Endothelial dysfunction contributes to hypertension, whereas hypertension also leads to endothelial dysfunction. In healthy endothelial tissues, a balance between endothelium-derived relaxing factors (EDRFs) and endothelium-derived contracting factors (EDCFs) is maintained. Endothelial dysfunction disturbs this balance. Several vasodilatory and vasoconstrictive factors regulate this balance. The endothelium secretes a number of vasodilator factors including NO. Generation of NO can activate guanylate cyclase (cGMP), which causes vasodilation through relaxation of vascular smooth muscle cells [56]. Another vasodilatory factor PGI_2 secreted by the endothelium inhibits platelet aggregation and vascular smooth muscle cell proliferation [57]. Endothelial cells also secrete several vascular contracting factors including angiotensin-II (Ang-II), endothelin-I (ET-I), dinucleotide uridine adenosine tetraphosphate (UP4A), and COX-derived TXA_2 [58]. Endothelins (ETs) are potent vasoconstrictor molecules having a key role in vascular homeostasis. Although there are three types of ET, vascular ECs mainly produce only ET-1, which has prominent roles in vasoconstriction. Active ET molecule is generated by the actions of an ET converting enzyme (ECE) found on the endothelial cell membranes. There are two basic types of ET-1 receptors: ET-A and ET-B, G-protein coupled receptors. Under normal conditions, the ET-A receptor is dominant in blood vessels [59]. ET-1 exerts vasoconstriction through activation of dihydropyridine channel or DHP channel or long lasting Ca^{++} channels (L-type) by binding to ET-A receptors on vascular smooth muscle cells. Smooth muscle cells expressed both ET-A and ET-B receptors. However, endothelial cells express only ET-B receptors, which negatively regulate NO release. Another vasorelaxation factor adenosine released from endothelial cells acts through purinergic receptor and maintains vascular perfusion [60]. Other than these factors, several cytokines and chemokines also play an important role in hypertension. Inflammatory cytokine induces generation of reactive oxygen species (ROS), one of the critical factors that link endothelial dysfunction and hypertension [61]. It is well established that Ang-II induces NADPH oxidases (NOX). But recent finding indicates additional source of ROS generation. In small subcutaneous arteries, a significant portion of Ang-II induced ROS is produced by COX-2. In the mouse aorta, the mitochondrial monoamine oxidase is another mediator of ROS generation and Ang-II or inflammation-induced endothelial dysfunction [62]. Therefore, mitochondrial monoamine oxidase-A and monoamine oxidase-B are also induced due to endothelial dysfunction in the vessels and generate a significant amount of H_2O_2 sufficient to quench endothelial NO. In spite of that, other mitochondrial ROS generating systems, that is, p66Shc, also contribute to hypertension-induced ROS production. ROS production is also regulated by several intracellular signaling, which further attenuate endothelial dysfunction and hypertension.

6. Endothelial dysfunction in heart failure

Heart failure (HF) is the most common cause of hospitalization in cardiovascular disease with a high mortality rate. Despite novel treatment options for patients suffering from HF, morbidity and mortality rates are still high. The impact of the growing HF on global public health is a great concern in health care research. With the advancement of medical management, survival of acute coronary disease and cardiac ischemia has been improved. However, in myocardial infarction, prognosis is still poor, as HF with preserved ejection fraction (HFpEF) has a 65% mortality rate at 5 years. While the heart as the failing "pumping" organ was an initial focus in research and treatment, neurohumoral activation and subsequently the role of a failing endothelium were recognized and investigated in recent years. Traditionally, HF was recognized as impairment of cardiac muscle activity, known as cardiomyopathy. Later, it was found that altered perfusion in cardiac arteries due to atherogenesis also contributes to cardiac ischemia that leads to cardiomyopathy. Reduced myocardial perfusion due to impaired ventricular function is at least in part a consequence of reduced endothelium-dependent vasodilator capacity of coronary arteries. The prominent regulatory activity of the vascular endothelium in HF was discovered about two decades ago, and its assessment in different cardiovascular disorders, including HF, has been the focus of intense research [63]. On the other hand, declined peripheral vasodilation causes higher systemic and pulmonary vascular resistance and together with stiffness of conductance arteries leads to increased afterload. Elevated afterload further increases cardiac workload and therefore worsens myocardial function. The decreased exercise capability is aggravated by vasomotor dysfunction of the skeletal muscle vessel by increases vascular resistance. Altered endothelial metabolism further contributes to increasing cardiac afterload [13]. Indeed, various aspects of endothelial function are affected in heart failure, including vasomotor, hemostatic, antioxidant, and anti-inflammatory activities [63, 64]. Differences also exist in the pattern of endothelial dysfunction depending on etiology, severity, and stability of HF in individual patients. Endothelial dysfunction also plays a central role in HF. The failing heart is characterized by an altered redox state with overproduction of ROS. The increasing evidence suggests that the abnormal cardiac and vascular phenotypes characterizing the failing heart are caused in large part by imbalances between NO bioavailability and oxidative stress [65]. During initial stages of HF, inflammatory mediators from the myocardium and altered local shear forces modulate gene expression, leukocyte infiltration, increased cytokine production, increased ROS generation, and diminished NO bioavailability. Clinical studies showed significant up-regulation of plasma markers of endothelial activation (e.g. E-selectin) and endothelial damage (e.g. vWF) in HF [22, 63]. However, it is difficult to determine if endothelial dysfunction is the cause or effect of the HF. Therefore, HF is regarded as thrombotic complication. As mentioned earlier, during atherogenesis, decreased lumen of cardiac arteries leads to reduced perfusion to the heart muscle. This phenomenon is coupled with increased sheer stress and impaired blood flow. This reduced perfusion either led to ischemia–reperfusion injury or coronary artery thrombosis [63]. Studies showed that endothelial dysfunction is one of the principle mediators of ischemia–reperfusion injury and thrombosis. This explains the increased endothelial dysfunction markers in coronary artery disease, HF, and thrombosis.

7. Endothelial dysfunction in stroke

The global burden of neurological diseases including cerebrovascular stroke has significantly increased, and development of new treatment modalities for cerebrovascular diseases is an urgent need. Cerebrovascular stroke can be broadly subdivided into acute ischemic stroke and hemorrhagic stroke [66]. Acute ischemic stroke is among the leading causes of death and long-term disability. Cerebrovascular stroke in small vessel has functional (lacunar stroke, cognitive impairment, gait, and movement disorders) and structural (small subcortical infarct, lacunar infarct, lacunes, white matter lesions, and micro bleeds) consequences. In the past few decades, the immense development of neuro-radiological methods enabled better imaging of cerebral blood vessels. From the clinical point of view, it is very important to identify the location of vascular lesion. However, the treatment strategies do not depend on the location of vascular impairment. It is now well recognized that endothelial dysfunction represents a systemic syndrome involving multiple vascular beds, including the cerebral vasculature [67]. Endothelial function is not uniform throughout the arterial system. It differs between organs and potentially also between different vascular beds within the same organ. Cerebral endothelium is probably one of the most specific types since it is the crucial element of the well-known blood-brain barrier (BBB). The BBB is a term used to describe the unique properties of the microvasculature of the central nervous system that protects the brain from harmful agents and pathogens [68]. CNS vessels are continuous non-fenestrated vessels, but also contain a series of additional properties that allow them to tightly regulate the movement of molecules, ions, and cells between the blood and the CNS. This heavily restricting barrier capacity allows BBB ECs to tightly regulate CNS homeostasis, which is critical to allow for proper neuronal function, as well as protect the CNS from toxins, pathogens, inflammation, injury, and disease. The cell-to-cell interaction with astrocytes, microglia, and neurons mainly played an important role for maintenance of BBB controlled by endothelial cells and pericytes [69].

However, the integrity of BBB is mainly disrupted due to decrease in endothelial cell-cell junction proteins and the detachment of pericytes from the endothelial membrane in homorganic condition [70]. Cerebral autoregulation maintains constant blood flow (CBF) through the brain in spite of changing mean arterial pressure. Autoregulation of cerebral blood flow consists of mechano- and chemoregulation. The serum level of carbon dioxide (CO_2) is directly controlled by the chemo-regulation independent of changes in mean arterial pressure [71]. However, mechano-regulation depends on transmural pressure gradient and endothelial vasodilatation.

8. Conclusion

From the above discussion, it is evident that CVDs and cardiovascular morbidity are associated with endothelial dysfunction, but the mechanistic links between inflammatory diseases, endothelial dysfunction, and CVDs have not been fully elucidated. The role of traditional cardiovascular risk factors in patients with inflammation, especially sterile inflammation, has received considerable attention, though traditional factors alone are insufficient to explain the excess burden of CVDs. It seems likely that sterile inflammation, a shared feature of CVDs, is involved in the pathogenesis of accelerated endothelial dysfunction.

Patients with chronic inflammatory/and or sterile inflammatory diseases are at high risk for cardiovascular morbidity and mortality. In many inflammatory diseases, this heightened risk of CVDs is reflected in early endothelial dysfunction, even in the absence of any other detectable diseases. Several others mechanisms, that is, auto-antibodies, oxidative stress, and interactions with traditional risk factors like dyslipidemia and insulin resistance might be involved. Therefore, further research in future is required to delineate the importance of these processes. So, the current approaches to diminish cardiovascular morbidity and mortality are focused on controlling traditional modifiable cardiovascular risk factors and reduction of disease risk. Therefore, the precise mechanisms leading to development of CVDs due to inflammation/or sterile inflammation need to explore. These studies might help to identify unique therapeutic targets to combat these diseases.

The endothelium therefore represents an integrator of vascular risk, and the study of its dysfunction may help elucidate mechanisms driving accelerated CVDs in future, which could help to develop therapeutic targets for control of CVDs.

Acknowledgements

This work was supported by grant Grant Challenges Canada (R-ST-POC-1807-13914) to GAK.

Author details

Indranil Biswas[1] and Gausal A. Khan[2]*

1 Cardiovascular Biology Research Program, Oklahoma Medical Research Foundation, Oklahoma City, OK, USA

2 Department of Physiology, Fiji School of Medicine, Fiji National University, Suva, Fiji Islands

*Address all correspondence to: gausalk@gmail.com

IntechOpen

© 2019 The Author(s). Licensee IntechOpen. This chapter is distributed under the terms of the Creative Commons Attribution License (http://creativecommons.org/licenses/by/3.0), which permits unrestricted use, distribution, and reproduction in any medium, provided the original work is properly cited.

References

[1] Krysko DV, Agostinis P, Krysko O, et al. Emerging role of damage-associated molecular patterns derived from mitochondria in inflammation. Trends in Immunology. 2011 April;**32**(4):157-164

[2] Deanfield JE, Halcox JP, Rabelink TJ. Endothelial function and dysfunction: Testing and clinical relevance. Circulation. 2007 March;**115**(10):1285-1295

[3] Widmer RJ, Lerman A. Endothelial dysfunction and cardiovascular disease. Global Cardiology Science & Practice. 2014;**2014**(3):291-308

[4] Cai H, Harrison DG. Endothelial dysfunction in cardiovascular diseases: The role of oxidant stress. Circulation Research. 2000 November;**87**(10):840-844

[5] Iantorno M, Campia U, Di DN, et al. Obesity, inflammation and endothelial dysfunction. Journal of Biological Regulators and Homeostatic Agents. 2014 April;**28**(2):169-176

[6] Vallance P, Chan N. Endothelial function and nitric oxide: Clinical relevance. Heart. 2001 March;**85**(3):342-350

[7] Zhao Y, Vanhoutte PM, Leung SW. Vascular nitric oxide: Beyond eNOS. Journal of Pharmacological Sciences. 2015 October;**129**(2):83-94

[8] Nishikimi T, Maeda N, Matsuoka H. The role of natriuretic peptides in cardioprotection. Cardiovascular Research. 2006 February;**69**(2):318-328

[9] Mitchell JA, Ali F, Bailey L, Moreno L, Harrington LS. Role of nitric oxide and prostacyclin as vasoactive hormones released by the endothelium. Experimental Physiology. 2008 January;**93**(1):141-147

[10] Smyth EM. Thromboxane and the thromboxane receptor in cardiovascular disease. Clinical Lipidology. 2010 April;**5**(2):209-219

[11] Qian H, Luo N, Chi Y. Aging-shifted prostaglandin profile in endothelium as a factor in cardiovascular disorders. Journal of Aging Research. 2012;**2012**:121390

[12] Hess CN, Kou R, Johnson RP, Li GK, Michel T. ADP signaling in vascular endothelial cells: ADP-dependent activation of the endothelial isoform of nitric-oxide synthase requires the expression but not the kinase activity of AMP-activated protein kinase. The Journal of Biological Chemistry. 2009 November;**284**(47):32209-32224

[13] Barthelmes J, Nagele MP, Ludovici V, Ruschitzka F, Sudano I, Flammer AJ. Endothelial dysfunction in cardiovascular disease and Flammer syndrome-similarities and differences. The EPMA Journal. 2017 June;**8**(2):99-109

[14] Cardaropoli S, Silvagno F, Morra E, Pescarmona GP, Todros T. Infectious and inflammatory stimuli decrease endothelial nitric oxide synthase activity in vitro. Journal of Hypertension. 2003 November;**21**(11):2103-2110

[15] Reyes LA, Boslett J, Varadharaj S, et al. Depletion of NADP(H) due to CD38 activation triggers endothelial dysfunction in the postischemic heart. Proceedings of the National Academy of Sciences of the United States of America. 2015 September;**112**(37):11648-11653

[16] Singh B, Biswas I, Garg I, et al. von Willebrand factor antagonizes nitric oxide synthase to promote insulin resistance during hypoxia. Biochemistry. 2014 January;**53**(1):115-126

[17] Matsubara K, Higaki T, Matsubara Y, Nawa A. Nitric oxide and reactive oxygen species in the pathogenesis of preeclampsia. International Journal of Molecular Sciences. 2015 March;**16**(3):4600-4614

[18] Chistiakov DA, Orekhov AN, Bobryshev YV. Endothelial barrier and its abnormalities in cardiovascular disease. Frontiers in Physiology. 2015;**6**:365

[19] Reitsma S, Slaaf DW, Vink H, van Zandvoort MA, Oude Egbrink MG. The endothelial glycocalyx: Composition, functions, and visualization. Pflügers Archiv. 2007 June;**454**(3):345-359

[20] Bonetti PO, Lerman LO, Lerman A. Endothelial dysfunction: A marker of atherosclerotic risk. Arteriosclerosis, Thrombosis, and Vascular Biology. 2003 February;**23**(2):168-175

[21] Oliver JJ, Webb DJ, Newby DE. Stimulated tissue plasminogen activator release as a marker of endothelial function in humans. Arteriosclerosis, Thrombosis, and Vascular Biology. 2005 December;**25**(12):2470-2479

[22] Yau JW, Teoh H, Verma S. Endothelial cell control of thrombosis. BMC Cardiovascular Disorders. 2015 October;**15**:130

[23] Ruggeri ZM, Mendolicchio GL. Adhesion mechanisms in platelet function. Circulation Research. 2007 June;**100**(12):1673-1685

[24] Gradinaru D, Borsa C, Ionescu C, Prada GI. Oxidized LDL and NO synthesis—Biomarkers of endothelial dysfunction and ageing. Mechanisms of Ageing and Development. 2015 November;**151**:101-113

[25] Badimon L, Padro T, Vilahur G. Atherosclerosis, platelets and thrombosis in acute ischaemic heart disease. European Heart Journal Acute Cardiovascular Care. 2012 April;**1**(1):60-74

[26] Andrassy M, Volz HC, Maack B, et al. HMGB1 is associated with atherosclerotic plaque composition and burden in patients with stable coronary artery disease. PLoS One. 2012;**7**(12):e52081

[27] Xu Q. Role of heat shock proteins in atherosclerosis. Arteriosclerosis, Thrombosis, and Vascular Biology. 2002 October;**22**(10):1547-1559

[28] Virmani R, Kolodgie FD, Burke AP, Farb A, Schwartz SM. Lessons from sudden coronary death: A comprehensive morphological classification scheme for atherosclerotic lesions. Arteriosclerosis, Thrombosis, and Vascular Biology. 2000 May; **20**(5):1262-1275

[29] Davignon J, Ganz P. Role of endothelial dysfunction in atherosclerosis. Circulation. 2004 June;**109**(23 Suppl 1):III27-III32

[30] Simionescu N, Vasile E, Lupu F, Popescu G, Simionescu M. Prelesional events in atherogenesis. Accumulation of extracellular cholesterol-rich liposomes in the arterial intima and cardiac valves of the hyperlipidemic rabbit. The American Journal of Pathology. 1986 April;**123**(1):109-125

[31] Ross R. The pathogenesis of atherosclerosis—An update. The New England Journal of Medicine. 1986 February;**314**(8):488-500

[32] Ross R. Atherosclerosis—An inflammatory disease. The New England Journal of Medicine. 1999 January;**340**(2):115-126

[33] Tabas I, Garcia-Cardena G, Owens GK. Recent insights into the cellular biology of atherosclerosis. The Journal of Cell Biology. 2015 April;**209**(1):13-22

[34] Burke AP, Kolodgie FD, Farb A, Weber D, Virmani R. Morphological predictors of arterial remodeling in coronary atherosclerosis. Circulation. 2002 January;**105**(3):297-303

[35] Guyton JR, Klemp KF. Development of the lipid-rich core in human atherosclerosis. Arteriosclerosis, Thrombosis, and Vascular Biology. 1996 January;**16**(1):4-11

[36] Tabas I, Lichtman AH. Monocyte-macrophages and T cells in atherosclerosis. Immunity. 2017 October;**47**(4):621-634

[37] Johnson JL. Metalloproteinases in atherosclerosis. European Journal of Pharmacology. 2017 December;**816**:93-106

[38] Cheng C, Tempel D, Van HR, et al. Shear stress-induced changes in atherosclerotic plaque composition are modulated by chemokines. The Journal of Clinical Investigation. 2007 March;**117**(3):616-626

[39] Miller YI. Toll-like receptors and atherosclerosis: Oxidized LDL as an endogenous toll-like receptor ligand. Future Cardiology. 2005 November;**1**(6):785-792

[40] Xie C, Ng H, Nagarajan S. OxLDL or TLR2-induced cytokine response is enhanced by oxLDL-independent novel domain on mouse CD36. Immunology Letters. 2011 June;**137**(1-2):15-27

[41] WoollardKJ, GeissmannF. Monocytes in atherosclerosis: Subsets and functions. Nature Reviews. Cardiology. 2010 February;7(2):77-86

[42] Pamukcu B, Lip GY, Devitt A, Griffiths H, Shantsila E. The role of monocytes in atherosclerotic coronary artery disease. Annals of Medicine. 2010 September;**42**(6):394-403

[43] Kobiyama K, Ley K. Atherosclerosis. Circulation Research. 2018 October;**123**(10):1118-1120

[44] Forrester SJ, Kikuchi DS, Hernandes MS, Xu Q, Griendling KK. Reactive oxygen species in metabolic and inflammatory signaling. Circulation Research. 2018 March;**122**(6):877-902

[45] Basu D, Hu Y, Huggins LA, et al. Novel reversible model of atherosclerosis and regression using oligonucleotide regulation of the LDL receptor. Circulation Research. 2018 February;**122**(4):560-567

[46] Mann JM, Davies MJ. Vulnerable plaque. Relation of characteristics to degree of stenosis in human coronary arteries. Circulation. 1996 September;**94**(5):928-931

[47] Schwartz SM, Galis ZS, Rosenfeld ME, Falk E. Plaque rupture in humans and mice. Arteriosclerosis, Thrombosis, and Vascular Biology. 2007 April;**27**(4):705-713

[48] Libby P. Mechanisms of acute coronary syndromes. The New England Journal of Medicine. 2013 August;**369**(9):883-884

[49] Quillard T, Araujo HA, Franck G, Shvartz E, Sukhova G, Libby P. TLR2 and neutrophils potentiate endothelial stress, apoptosis and detachment: Implications for superficial erosion. European Heart Journal. 2015 June; **36**(22):1394-1404

[50] Libby P. How does lipid lowering prevent coronary events? New insights from human imaging trials. European Heart Journal. 2015 February;**36**(8):472-474

[51] Back M, Ketelhuth DF, Agewall S. Matrix metalloproteinases in atherothrombosis. Progress in Cardiovascular Diseases. 2010 March;**52**(5):410-428

[52] Reininger AJ, Bernlochner I, Penz SM, et al. A 2-step mechanism of arterial thrombus formation induced by human atherosclerotic plaques. Journal of the American College of Cardiology. 2010 March;**55**(11):1147-1158

[53] Insull W Jr. The pathology of atherosclerosis: Plaque development and plaque responses to medical treatment. The American Journal of Medicine. 2009 January;**122**(1 Suppl):S3-S14

[54] Brandes RP. Endothelial dysfunction and hypertension. Hypertension. 2014 November;**64**(5):924-928

[55] Bleakley C, Hamilton PK, Pumb R, Harbinson M, McVeigh GE. Endothelial function in hypertension: Victim or culprit? Journal of Clinical Hypertension (Greenwich). 2015 August;**17**(8):651-654

[56] Munzel T, Daiber A, Ullrich V, Mulsch A. Vascular consequences of endothelial nitric oxide synthase uncoupling for the activity and expression of the soluble guanylyl cyclase and the cGMP-dependent protein kinase. Arteriosclerosis, Thrombosis, and Vascular Biology. 2005 August;**25**(8):1551-1557

[57] Soler M, Camacho M, Escudero JR, Iniguez MA, Vila L. Human vascular smooth muscle cells but not endothelial cells express prostaglandin E synthase. Circulation Research. 2000 September;**87**(6):504-507

[58] Ashton AW, Ware JA. Thromboxane A2 receptor signaling inhibits vascular endothelial growth factor-induced endothelial cell differentiation and migration. Circulation Research. 2004 August;**95**(4):372-379

[59] Rodriguez-Vita J, Ruiz-Ortega M, Ruperez M, et al. Endothelin-1, via ETA receptor and independently of transforming growth factor-beta, increases the connective tissue growth factor in vascular smooth muscle cells. Circulation Research. 2005 July;**97**(2):125-134

[60] Bohm F, Ahlborg G, Johansson BL, Hansson LO, Pernow J. Combined endothelin receptor blockade evokes enhanced vasodilatation in patients with atherosclerosis. Arteriosclerosis, Thrombosis, and Vascular Biology. 2002 April;**22**(4):674-679

[61] Schulz E, Gori T, Munzel T. Oxidative stress and endothelial dysfunction in hypertension. Hypertension Research. 2011 June;**34**(6):665-673

[62] Dikalov SI, Ungvari Z. Role of mitochondrial oxidative stress in hypertension. American Journal of Physiology. Heart and Circulatory Physiology. 2013 November;**305**(10):H1417-H1427

[63] Bauersachs J, Widder JD. Endothelial dysfunction in heart failure. Pharmacological Reports. 2008 January;**60**(1):119-126

[64] Mann DL. Inflammatory mediators and the failing heart: Past, present, and the foreseeable future. Circulation Research. 2002 November;**91**(11):988-998

[65] Marti CN, Gheorghiade M, Kalogeropoulos AP, Georgiopoulou VV, Quyyumi AA, Butler J. Endothelial dysfunction, arterial stiffness, and heart failure. Journal of the American College of Cardiology. 2012 October;**60**(16, 16): 1455-1469

[66] AndersenKK,OlsenTS,DehlendorffC, Kammersgaard LP. Hemorrhagic and ischemic strokes compared: Stroke severity, mortality, and risk factors. Stroke. 2009 June;**40**(6):2068-2072

[67] Ludewig P, Winneberger J, Magnus T. The cerebral endothelial cell as a key regulator of inflammatory processes in sterile inflammation.

Journal of Neuroimmunology. 2019 January;**326**:38-44

[68] Hawkins BT, Egleton RD. Pathophysiology of the blood-brain barrier: Animal models and methods. Current Topics in Developmental Biology. 2008;**80**:277-309

[69] Yang S, Jin H, Zhu Y, et al. Diverse functions and mechanisms of Pericytes in ischemic stroke. Current Neuropharmacology. 2017;**15**(6):892-905

[70] Hsu MJ, Hsu CY, Chen BC, Chen MC, Ou G, Lin CH. Apoptosis signal-regulating kinase 1 in amyloid beta peptide-induced cerebral endothelial cell apoptosis. The Journal of Neuroscience. 2007 May;**27**(21):5719-5729

[71] Karnushina IL, Spatz M, Bembry J. Cerebral endothelial cell culture. I. The presence of beta 2 and alpha 2-adrenergic receptors linked to adenylate cyclase activity. Life Sciences. 1982 March;**30**(10):849-858

Chapter 6

Association between Peripheral Blood Inflammatory Markers, Endothelial Dysfunction Markers, and Depression

Olga Vladimirovna Vorob'eva,
Victoria Vyacheslavovna Fateeva,
Ksenia Vladimirovna Nikulina,
Kristina Konstantinovna Khacheva,
Gulnara Rinatovna Khakimova and Oleg Ilyich Epstein

Abstract

The authors present an analysis of current research and their own data on the link between endothelial dysfunction (ED) and the severity of depression in middle-aged patients with cerebral microangiopathy. Levels of peripheral inflammatory and endothelial dysfunction markers were measured using the enzyme-linked immunosorbent assay (ELISA). The results of the comparative and correlation analyses showed a statistically significant correlation between the severity of depression and increased levels of inflammatory, as well as endothelial dysfunction

Keywords: inflammatory marker, endothelial dysfunction maker, depression, cerebral microangiopathy, Divaza

1. Introduction

Depression and cardiovascular diseases (CVD) are the leading problems of modern healthcare [1]. The global prevalence of depressive disorders ranges from 4.4 to 20%. About 350 million people suffer from depression [2, 3].

The World Health Organization (WHO) considers depression and atherosclerosis as the main causes of disability in CVD and predicts that by 2030, depression will be the leading one. The fact that a depressive disorder is more common in women and the elderly people (in whom CVD is often underestimated) determines the need for effective preventive measures in these risk groups [4, 5].

The multiple factors contributing to the strong association between depression and atherosclerosis are not well studied. Authors discuss various mechanisms of linkage between these diseases [6–12], i.e., the activation of neuroendocrine system [5].

Frequently, depression is considered an unfavorable prognostic factor for the CVD development [6–8]. Unhealthy lifestyle (physical inactivity, smoking), as well

 IntechOpen

as inflammation and metabolic syndrome, has been proposed to explain the relationship between depression and the development of atherosclerosis [9, 10, 12, 13].

In patients with depression, elevated levels of such markers of systemic inflammation as pro-inflammatory cytokines (tumor necrosis factor alpha (TNF-α), interleukin-6 (IL-6)), C-reactive protein (CRP), as well as reduced levels of interleukin-1 (IL-1) and interleukin-8 (IL-8) [14–17], lead to hyperactivation of the hypothalamic-pituitary-adrenal axis [18]. Also intestinal inflammation causes remote effect on the brain and depression. Unbalanced gut microbiota shifts tryptophan metabolism toward kynurenine pathway affecting gamma-aminobutyric acid (GABA), dopamine, and serotonin levels [19, 20]. Studies have shown that stimulation of the vagus nerve modulates the inflammatory response and relieves the depression symptoms [21, 22].

Many studies focus on the activation of inflammatory processes as a common pathogenetic pathway between CVD, cerebrovascular disorders, and depression. In this respect, there are two aspects of the role of pro-inflammatory mechanisms: the first reveals the causal relationship between depression and inflammation and/or endothelial dysfunction (ED); the second shows the connection between inflammation, ED, and atherosclerosis [5]. ED and/or elevated levels of peripheral markers of inflammation are found in atherosclerosis and cardiovascular events [23]. Patients at high risk of acute coronary syndrome can be identified based on the elevated levels of inflammatory markers [24]. These markers are associated with atherosclerosis and activation of the immune system. The immune system components can indirectly influence the progression of atherosclerotic vascular lesions.

Risk factors for CVD, which at the same time predispose to atherosclerosis, cause dysfunction of the vascular endothelium. The vascular endothelium, as a regulator of vascular homeostasis, is involved in different processes, such as keeping a balance between vasodilatation and vasoconstriction, inhibition and stimulation of smooth muscle cell proliferation and migration, thrombus formation, and fibrinolysis [25].

When imbalance occurs, ED develops, causing excessive vascular permeability, platelet aggregation, leukocyte adhesion, and production of cytokines. A failure in synthesis or a violation of the activity of nitric oxide (NO) (which is manifested by insufficient vasodilatation) also is considered as a potential mechanism by which endothelial dysfunction leads to atherosclerosis. The low systemic levels of NO have been postulated to be responsible for the increased risk of cardiovascular events observed in subjects with depression, as NO produces vasodilatation [26].

Among other cardiovascular risk factors, obesity increases morbidity and mortality from CVD. The relationship between depression and obesity is complex and includes behavioral, biological, and pathological associations [27]. The mechanisms leading to the development of obesity affect the vascular system and especially the endothelium function, mainly due to the pathological role of TNF-α in the generation of reactive oxygen species and reducing the availability of NO [28].

Significantly more research is needed to further study the multiple associations between inflammation, markers of ED, and depression in patients with CVD [29]. Therefore, we have performed our own study aimed at identifying the relationship between the levels of peripheral inflammatory markers and markers of ED with the level of depression in patients with cerebral microangiopathy associated with arterial hypertension and cerebral atherosclerosis.

2. Materials and methods, results, discussion

The study included middle-aged (44–60 years old) outpatients of both sexes, with cerebral microangiopathy. The diagnosis was based on (1) the presence of a vascular disease (arterial hypertension and/or cerebral atherosclerosis) and focal neurological symptoms in combination with cerebral symptoms (headache, dizziness, noise in the head, memory loss, etc.), (2) ultrasonic signs of cerebrovascular atherosclerosis according to duplex scanning of the main arteries of the head, and (3) signs of morphological changes in the brain according to neuroimaging data (subcortical and periventricular leucoaraiosis and/or focal changes of gray and white matter in the form of postischemic cysts and/or lacunary strokes and/or diffuse atrophic changes in the form of ventricular dilatation or subarachnoid spaces). Also all patients had depression (≥8 points on the Hospital Anxiety and Depression Scale (HADS)). Study participants were capable of communicating with the researcher, understanding and signing the informed consent form.

All patients received Divaza for 3 months. Divaza is a drug with endothelial protective as well as nootropic (neuroprotective, neurotrophic) activity and good safety profile [30–54].

Combination drug Divaza was developed based on previously discovered pharmacological effects of its active pharmaceutical ingredients—release-active antibodies to S100 (RAF of Abs to S100) and release-active antibodies to endothelial NO synthase (RAF of Abs to eNOS) (including their different technological versions). The important feature of RA forms of a substance is the ability to exert modifying effect on the original substance or its target [Symmetry, 2018]. Since this activity is released during technological processing of the original substance, it has been termed as the released activity and the produced drug as the released-active drug.

Preclinical studies have shown that Divaza successfully combines properties of the monocomponents [see ***Appendix 1****]. The overall therapeutic effect results in an enhanced ability of experimental animals to remember and reproduce the developed skills as well as overcome artificially induced anxiety and depression. Furthermore, therapeutic effect of Divaza on functional activity of endothelium has been confirmed by normalizing such integrative indicator of the physiological status as blood pressure: RAF of Abs to eNOS exerted hypotensive effect in animals with arterial hypertension and, at the same time, did not alter parameters of system hemodynamics in normotensive condition. Importantly, RAF of Abs to eNOS given to animals along with nitroglycerin, an exogenous NO-donor, resulted neither in a drug-drug interaction nor in excessive vasodilation (the drug did not enhance the nitroglycerin effect in normotensive animals).*

Noteworthy, Divaza, while exerting a normalizing psychotropic effect, did not cause sedation and/or muscle relaxation (unpublished data).

RAF of Abs to S100, RAF of Abs to eNOS, and combination drug Divaza did not exert any toxic effect as assessed by evaluation of single- and repeat-dose (chronic) toxicity, genotoxicity, reproductive and developmental toxicity, local tolerance, allergenic properties and immunotoxicity even when administered at doses more than 100 times exceeding the recommended human daily doses [31, 32].

During the study, patients were allowed to take concomitant drugs for correcting the main risk factors for CVD development (antihypertensive drugs, diuretics, antiplatelet drugs, anticoagulants, statins, and antidiabetic drugs).

An assessment of the dynamics of clinical and laboratory parameters was performed at baseline and after 3 months of therapy. Clinical examination of patients included collecting and analyzing complaints, anamnesis, anthropometric measurements of body weight and height to calculate body mass index (BMI), and thorough neurological examination, including neuropsychological testing and emotional status assessment using HADS.

Serum levels of peripheral markers of endothelial inflammation (CRP, monocyte chemoattractant protein-1 (MCP-1)) as well as biomarkers of ED (endothelin-1, eNOS, vascular endothelial growth factor (VEGF), S100B protein, von Willebrand factor, fibrinogen) were determined by ELISA. The amount of circulating desquamated endothelial cells (CECs) was detected according to J. Hladovec with modifications.

ELISA reference values of biomarkers were CRP up to 1 mg/l, MCP-1 0.228–0.475 ng/ml, endothelin-1 0.3–0.5 ng/ml, eNOS y < 450 units VEGF 101–409 pg/ml, S100B protein 29–56.5 μg/l, CECs 2–4 cells/100 μl, von Willebrand factor < 4 cu, and fibrinogen 2–4 g/l.

The study included a total of 262 patients between 44 and 60 years of age (mean age 54.2 ± 7) with cerebral microangiopathy and depression. Among patients there were 110 (42.1%) men (mean age 50.8 ± 6.3) and 152 (57.9%) women (mean age 57.4 ± 8.1). One hundred ninety-eight (75.6%) patients had grade 1 arterial hypertension, 64 (24.4%) had grade 2 arterial hypertension, 152 (58%) had type 2 diabetes mellitus, and 169 (64.5%) had obesity (BMI > 30 kg/m^2). The severity of depression on the HADS hospital scale was 12.84 ± 5.67 points.

The sociodemographic and clinical characteristics of patients are presented in **Table 1**.

The initial patient sample (n = 262) was divided into two groups: a group with clinically significant depression (≥11 points on the HADS scale) (n = 146) and a group with subclinically significant depression (8–10 points on the HADS scale) (n = 116). At the time of inclusion in the study, no statistically significant differences between groups by sociodemographic characteristics were identified.

A comparative analysis revealed that in the group of patients with clinically significant depression (n = 146), the CRP level was 6.11 mg/l, which was statistically different from the group of patients with subclinical depression (n = 116) where CRP was 2.03 mg/L ($p < 0.05$). The level of MCP-1 in the group of patients with clinically significant depression (n = 146) was 2.02 ng/ml, while in the group of patients with subclinical depression (n = 116), it was 0.66 ng/ml ($p < 0.05$).

After a correlation analysis, the presence of a statistically significant direct linear association between the severity of depression and the level of CRP (correlation coefficient r = 0.85, $p < 0.05$) as well as between the severity of depression and the level of MCP-1 (r = 0.8, $p < 0.05$) was shown.

It was revealed that elevated values of peripheral markers of endothelial inflammation are associated with clinically significant depression. The results are presented in **Table 2**. A reliable association was determined between inflammation and depression ($p < 0.001$), which remained significant after adjusting for risk factors for CVD.

Parameter	Main group (n=262)
Mean age, years:*	54,2±7
Sex, %	
Men	42,1
Women	57,9
BMI, %	
>30 kg/m^2	64,5
≤30 kg/m^2	35,5
Comorbidities, %	
Arterial hypertension I/II grade	75,6/24,4
Type 2 diabetes mellitus	58,0
The mean HADS score :*	12,84±5,67

*Note:*data shown as M±SD; BMI – Body Mass Index, HADS – Hospital Anxiety and Depression Scale*

Table 1.
Baseline clinical characteristics of patients.

After 12 weeks of treatment with Divaza, we observed a statistically significant change in the values of peripheral inflammatory markers toward the reference values: before treatment, the mean CRP value was 9.25 ± 3.5 mg/l, and after treatment it was 7.73 ± 3.12 mg/l ($p < 0.01$); before treatment the average value of MCP-1 was 2.81 ± 1.3 ng/ml, and after treatment it became 1.9 ± 1.2 ng/ml ($p < 0.01$).

Depending on the individual dynamics of peripheral inflammatory marker values, all patients (n = 262) were divided into two groups: with positive (n = 176) and with negative (n = 45) dynamics. The values of peripheral markers of endothelial inflammation in the groups before and after treatment are presented in **Table 3**.

Normalization of peripheral inflammatory and endothelial dysfunction markers led to psycho-emotional status improvement in patients (**Table 4**).

The results show that the reduction of inflammation was accompanied by a decrease in the severity of depressive symptoms, which, in turn, confirms the association between inflammation and depression.

Markers	OR*	95% CI	p	OR**	95% CI	p
CRP	4,17	1,18-14,7	0,001	1,5	1,3-1,8	0,001
MCP-1	4,67	1,99-10,97	0,001	1,67	0,63-2,86	0,001

*Note: # corrected for age, type 2 diabetes, obesity * excluding adjustment for age, diabetes mellitus, obesity; ** adjusted for age, diabetes mellitus, obesity; CRP – C-reactive Protein, MCP-1 – Monocyte Chemoattractant Protein 1; OR– Odds Ratio, CI – Confidence Interval*

Table 2.
Association between peripheral markers of endothelial inflammation and depression#.

Markers	Baseline		After Divaza treatment	
	Group with individual positive dynamics of endothelial inflammation marker values (n=176)	Group with individual negative dynamics of endothelial inflammation marker values (n=45)	Group with individual positive dynamics of endothelial inflammation marker values (n=176)	Group with individual negative dynamics of endothelial inflammation marker values (n=45)
CRP, mg/l	4,1±1,8	6,8±3,2	3,75±1,5*	5,2±2,7*
MCP-1, ng/ml	1,02±0,53	2,03±0,97	0,55±0,28*	1,5±0,63*

*Note: Data shown as M±SD; *p<0.05 between the groups before and after treatment, CRP – C-reactive Protein, MCP-1 – Monocyte Chemoattractant Protein 1*

Table 3.
Peripheral markers of endothelial inflammation level before and after treatment with Divaza.

The severity of depression before and after Divaza treatment

Baseline		After Divaza treatment	
Group with individual positive dynamics of endothelial inflammation marker values (n=176)	Group with individual negative dynamics of endothelial inflammation marker values (n=45)	Group with individual positive dynamics of endothelial inflammation marker values (n=176)	Group with individual negative dynamics of endothelial inflammation marker values (n=45)
13,7±6,8	12,85±6,3	8,12±4,1*	11,6±5,7

*Note: *p<0.05 between the groups before and after treatment*

Table 4.
The severity of depression before and after Divaza treatment.

To assess the possible involvement of other endothelial mediators in the development of depression in patients with cerebral microangiopathy, an integrative endothelial dysfunction index was calculated.

In factor analysis the most significant variables involved in the formation of the integrative endothelial dysfunction index were identified. The levels of laboratory markers of ED with evidence or possible role in the development of endothelial dysfunction, CRP, MCP-1, VEGF, fibrinogen, Willebrand factor, S100 protein, CECs, eNOS activity, and endothelin-1, were considered as variables.

CRP, MCP-1, endothelin-1, CECs, and fibrinogen levels statistically significantly formed one active factor—the integrative endothelial dysfunction index.

Using group of patients who had all laboratory values of ED within the reference range (n = 41), we defined an average value of the integrative endothelial dysfunction index equal to 1.32 ± 0.38 conv. units as the age norm.

Comparative analysis of the integrative endothelial dysfunction index values revealed its statistically significant increase in the group of patients with clinically significant depression (0.12 ± 0.04 conv. units) compared with the group of patients with subclinical depression (1.14 ± 0.3 conv. units) ($p < 0.05$).

A correlation analysis identified the presence of an inverse statistically significant correlation between the integrative endothelial dysfunction index and severity of depression ($r = -0.83$; $p < 0.05$).

When evaluating associations, we found that the integrative endothelial dysfunction index is associated with clinically significant depression (OR 1.36, 95% CI 1.23–1.45). The data are presented in **Table 5**.

The main hypothesis of this study was that in middle-aged patients with cerebral microangiopathy, the presence of depression is associated with modified levels of both peripheral inflammatory and endothelial dysfunction markers. To the best of our knowledge, patients with clinically significant depression were more likely to have elevated values of peripheral markers associated with endothelial inflammation and modified levels of ED markers compared with patients with subclinical depression ($p < 0.05$).

When studying the correlation between the severity of depression and elevated values of peripheral markers of inflammation, as well as altered levels of markers of ED, a significant correlation was found; furthermore there were no association with age, obesity, and diabetes mellitus ($p < 0.001$).

The concomitant administration of drugs aimed to correcting the main risk factors of CVD with Divaza helped to normalize the levels of peripheral inflammatory markers ($p < 0.05$), which was accompanied by the reduction of depressive symptoms in middle-aged patients with cerebral microangiopathy.

Integrative endothelial dysfunction index#

OR*	95% CI	p	OR **	95% CI	p
1,91	1,55-2,36	0,001	1,36	1,23-1,45	0,001

Note: # adjusted for age, type 2 diabetes, obesity; * excluding adjustment for age, diabetes mellitus, obesity; ** adjusted for age, diabetes mellitus, obesity; CRP – C-reactive Protein, MCP-1 – Monocyte Chemoattractant Protein 1; OR– Odds Ratio, CI – Confidence Interval

Table 5.
Integrative endothelial dysfunction index#.

The study revealed the association between abnormal levels of peripheral markers of inflammation, markers of ED, and the severity of depression in middle-aged patients with cerebral microangiopathy accompanied by arterial hypertension and cerebral atherosclerosis. Drug therapy of endothelial dysfunction contributed to the normalization of the emotional status in these patients.

Inflammation can be considered as a causative agent of the development of depression, which is associated with the brain vascular network damage. Damage of the endothelium stimulates the release of pro-inflammatory cytokines, which play a leading role in the deposition of lipids in the arterial wall and proliferation and migration of smooth myocytes into the intima of the blood vessels, causing thrombus formation followed by occlusion of the blood vessels.

Previous studies showed that patients with severe types of depression had elevated levels of peripheral inflammatory markers compared with patients without depression [54]. In particular, a major depressive disorder is accompanied by the dysregulation of the immune system with activation of the factors participating in the inflammatory response [55]. An increase in the secretion of pro-inflammatory cytokines (IL-1β, IL-2, IL-6, Interferon-gamma, TNF-α, etc.) and CRP has been demonstrated [56]. Our results further indicate the dependence of the level of depression on the levels of inflammatory markers.

As a possible mechanism of the association between peripheral inflammatory markers and depression, the involvement of microglia, which is able to use pro-inflammatory cytokines as catalysts for tryptophan degradation, was considered. Alteration of tryptophan metabolism can cause depressive symptoms by reducing serotonin synthesis [57].

IL-6 is a potent stimulator of the hypothalamic-pituitary-adrenal axis and induces the release of other pro-inflammatory cytokines [58]. These observations led Leonard B.E. to the conclusion that depression is a disease of inflammation in response to chronic psychological stress. The author considered stress as a common denominator in the etiopathology of cerebrovascular diseases, cardiovascular diseases, and depression [59].

CRP is a protein of the acute phase of inflammation that is activated by pro-inflammatory cytokines (e.g., IL-6) [60]. CRP is usually studied in the context of vascular medicine; however, elevated CRP levels are a risk factor for atherosclerosis and cardiovascular diseases. Therefore, the relationship between CRP and cardiovascular diseases was beyond the scope of cardiology. In recent decades, CRP has become a central element for psychiatric research, especially in the area of depression. Depressed patients have been shown to exhibit significantly higher levels of CRP compared to patients in the control group [61].

Wium-Andersen et al. demonstrated an association between CRP levels and psychological stress in 73,131 Copenhagen residents. After adjusting for age, gender, smoking, physical activity, and chronic somatic diseases, it was noted that elevated levels of CRP correlated with depression [62].

The meta-analysis performed by Howren et al. summarizes data from studies on participants with elevated level of inflammatory markers (CRP, IL-1, IL-6) and depression, taking into account the BMI. Researchers concluded that there are associations between high CRP level and depression after adjusting for BMI [63]. This suggests that the relationship between CRP and depressive disorders does not depend on BMI. In our study, similar results were obtained: an association was found between the level of CRP and depression—OR 4.17 (95% CI 1.18–14.7, $p < 0.001$), which persisted after adjustment for age, obesity, and diabetes mellitus —OR 1.5 (95% CI 1.3–1.8, $p < 0.001$). These facts demonstrate the independent character of the mutual effects between the activity of the inflammatory processes and the level of depression.

Despite this, many hypotheses have been proposed regarding the association of elevated CRP values with depression using various covariates that can affect these relationships: from BMI and general physical health to the type of depression, which makes it likely that there is a multifactorial and bi-directional relationship between depression and inflammation [64].

Frasure-Smith et al. demonstrated that the association between CRP level and depression is important for predicting future adverse cardiovascular events [65]. There is a lot of evidence in the scientific literature about significant and sustained elevations of CRP level in depressed patients, which may or may not normalize after remission of depression symptoms, indicating a continued possibility of adverse cardiovascular events even after recovery from depression [66].

In 2015, the results of Setiawan E. et al. were published, where authors concluded that depression causes inflammation. The researchers obtained the first nonexperimental evidence that besides neurons, glial cells, namely, microglia, play an important role in depressive disorders [67]. However, despite these findings, it is difficult to conclude that depressive disorders are a trigger of inflammation. It is possible that stress and other risk factors lead to depressive disorders that affect the activation of microglia and a change in its structure, in turn, increasing the level of depression, i.e., it may be a continuum. The data obtained in the present study on the reduction of the level of depression ($p < 0.05$) accompanied by a decrease in the inflammatory activity of the endothelium with the help of a therapy that does not have an antidepressant effect also suggest the two-sided associations between inflammation and depression.

ED is a vascular phenotype that predisposes to atheromatosis and atherosclerosis and, therefore, may be a predictor of cardiovascular events [68].

Endothelial dysfunction can lead to immunological changes, activation, adhesion of white blood cells, and aggregation of platelets in the vascular damage area. The attachment of monocytes and lymphocytes to endothelial cells is associated with the activation of cell adhesion molecules [69]. Chronic mild inflammation is known as a predictor of myocardial infarction and ischemic stroke. ED is a "critical intermediate phenotype" in the association between mild inflammation and CVD. ED can be considered as an "intermediate phenotype" in depression based on the presence of mild chronic inflammation in many patients with depressive symptoms [70]. Depression can be regarded as a chronic stressor that contributes to the development of ED due to the disruption of cell adhesion, platelet hypercoagulation. Depression is associated with higher levels of MCP-1, p-selectin, and others. Some researchers consider ED as a biomarker of arterial atheromatosis, which can be a sign of depression [71]. The present study has shown that there are strong associations between inflammatory markers and markers of proatherogenic activity (endothelin-1, CECs, and fibrinogen). Based on the values of CRP, MCP-1, endothelin-1, CECs, and fibrinogen, an integrative indicator of endothelial dysfunction had higher values in patients with clinically significant depression, compared with patients with subclinical depression ($p < 0.01$). An inverse statistically significant correlation was found between the integrative endothelial dysfunction index and severity of depression ($r = -0.83$; $p < 0.05$). Moreover, this association between the integrative endothelial dysfunction index and depression did not depend on age, type 2 diabetes mellitus, and obesity ($p < 0.001$)—the risk factors known to raise the incidence of CVD.

Do et al. focused on the association of individual symptoms of depression (a symptom of hopelessness that can turn into suicidality) with markers of ED.

The researchers concluded that negative psychosocial features which can affect cardiovascular diseases in part through their impact on the early stages of atherosclerosis, as well as specific psychosocial features such as hopelessness, may play a more significant role in this process than other depressive symptoms [72].

Thus, ED is a crucial factor in the bilateral relationship between depression, chronic inflammation, and cardiovascular diseases [73].

3. Conclusions

Despite growing evidence that underlines the bilateral relationship between depression and CVD and the fact that the mechanisms of their connection were partially identified (e.g., inflammation, ED), further research with a large sample size is required.

Novel pharmacological approaches based on discoveries related to the immune and neurotransmitter systems are in high demand.

The development and implementation of preventive measures and lifestyle correction, which can reduce the burden of cerebrovascular diseases and depression, are required as well.

Conflict of interest

The studies of the combination of released-active form of antibodies to S100 protein and released-active form of antibodies to endothelial NO synthase, mentioned in this review, were funded by a grant from OOO "NPF 'MATERIA MEDICA HOLDING'." Fateeva V.V., Nikulina K.V., Khacheva K.K., Khakimova G.R. and Epstein O.I. are employees and a founder of the OOO "NPF 'MATERIA MEDICA HOLDING'," respectively. Divaza is a preparation manufactured and marked by OOO "NPF 'MATERIA MEDICA HOLDING'." Patents on Divaza belong to Epstein O.I. Vorob'eva O.V. received an investigator grant from OOO "NPF 'MATERIA MEDICA HOLDING'" to conduct the clinical trials of Divaza, mentioned in this review.

Abbreviations

CVD	cardiovascular diseases
CRP	C-reactive protein
TNF-α	tumor necrosis factor alpha
IL	interleukin
GABA	gamma-aminobutyric acid
ED	endothelial dysfunction
HADS	hospital anxiety and depression scale
RA	release-activity
RAF of Abs to S100	release-active form of antibodies to S100B
RAF of Abs to eNOS	release-active form of antibodies to endothelial NO synthase
BMI	body mass index
ELISA	enzyme-linked immunosorbent assay
MCP-1	monocyte chemoattractant protein-1
CEC	circulating endothelial cells

Appendix

Drug	Studied activity	Test system	Results	Ref
Pharmacodynamics				
RAF of Abs to S100	Neuroprotective activity	Mouse neuroblastoma C-1300 cells deprived of glucose and O2	RAF of Abs to S100 added to the incubation medium accurately before or 20 h before the hypoxia induction increased the number of survived cells by 1.8- and 2.1-times, respectively	[33]
	Antidepressive activity	Adult outbred male albino rats	RAF of Abs to S100 increased the number of the wheel turns in Nomura forced swimming test by 1.8-2.2-times	[34, 35, 36]
			RAF of Abs to S100 decreased the duration of immobility in Porsolt forced swimming test by 1.6-times	[35]
	Anxiolytic activity	Adult outbred male albino rats	RAF of Abs to S100 increased the number of punished water intakes in Vogel conflict test by 1.4-3.2-times	[34, 36, 37, 38, 39]
			RAF of Abs to S100 increased the number of entries into open arms of EPM, the time spent in open arms and the number of leanings over the edge of the maze by 1.9-, 5.4- and 4.9-times, respectively	[34, 39]
			RAF of Abs to S100 increased the number of entries into the center of the open field up to 2.4±0.7 vs 0±0 in control group	[34, 39]
		Adult Rj:Wistar (Han) male rats	RAF of Abs to S100 increased the number of punished water intakes in Vogel conflict test by 1.5-times	[38]
RAF of Abs to eNOS	Endothelio-protective activity	Adult Wistar male rats with NO deficiency, induced by L-NAME	RAF of Abs to eNOS reduced arterial blood pressure (184.3±7.0 mm Hg vs 190.3±6.7 mm Hg in L-NAME-group) and the exhaustion of myocardial fractional flow reserve by 11%	[40]
		Adult Wistar male rats with NO deficiency, induced by L-NAME	RAF of Abs to eNOS improved microcirculation in the ischemic area, stimulated neoangiogenesis and promoted recruitment of additional capillaries into general circulation; RAF of Abs to eNOS improved the metabolism of endothelial capillaries and significantly	[41]

Drug	Studied activity	Test system	Results	Ref
			decreased the number of desquamated endotheliocytes that was the unique morphological criteria for endothelium damage degree	
		Adult Wistar male rats with hypoestrogen-induced NO deficiency	RAF of Abs to eNOS reduced arterial blood pressure (158.5±15.0 mm Hg vs 160.0±6.2 mm Hg in control group) and the exhaustion of myocardial fractional flow reserve by 26.9%	[40]
	Antidepressive activity	Adult outbred male rats	RAF of Abs to eNOS increased the number of punished water intakes in Vogel conflict test by 1.6-times	[42]
	Anxiolytic activity	Adult outbred male rats	RAF of Abs to eNOS increased the number of entries into open arms of EPM and time spent in open arms up to 1.50±0.71 and 34.4 8.50 s (vs 0±0 in control group for both parameters), respectively	[42]
	Influence on the cardiovascular system	Hypertensive adult ISIAH male rats	RAF of Abs to eNOS reduced arterial blood pressure by 5.7%	[43]
		Normotensive adult Wistar male rats	RAF of Abs to eNOS did not affect systemic hemodynamics, did not augment nitroglycerine effects	[44]
Divaza	Endothelio-protective activity	Thoracic aorta rings of adult male SHR rats	Divaza improved acetylcholine-induced relaxation of aorta vascular smooth muscle by 1.4-times	Un-published data
	Neuroprotective (antioxidant) activity	Adult outbred male rats with experimental acute hemic hypoxia	Divaza decreased the content of diene conjugates in the cerebral hemispheres by 9.7-27.8% in the heptane fraction and by 7.5-47.4% in isopropanol fraction. The accumulation of 2-thiobarbituric acid-reactive products was reduced by 20.1-27.5%	[45]
	Neurotropic activity (neurite outgrowth)	Brain cortex neurites of rat fetuses (gestational day 17) without any branching	Divaza increased the value of neurite outgrowth by 9.6%	Un-published data
	Anti-amnestic activity	Adult Wistar male rats with β-amyloid-induced amnesia	Divaza increased the latency of the entry into the dark compartment of CPAR experimental chamber by 3-times	[46]
	Antidepressive activity	Adult outbred male rats	Divaza increased the number of the wheel turns in Nomura	[47]

Drug	Studied activity	Test system	Results	Ref
			forced swimming test by 1.8-times	
	Anxiolytic activity	Adult outbred male rats	Divaza increased the number of punished water intakes in Vogel conflict test by 2.5-times	[47]
Mechanisms of action				
RAF of Abs to S100	Influence on Abs to S100 activity (ability to induce LTP)	Hippocampal slices (400 μm) of mature Wistar rats	Abs to S100 inhibited the induction of LPT, whereas RAF of Abs to S100 offset this inhibiting activity	[48]
	Influence on electrical properties of cell membranes	Isolated neurons of Helix pomatia	RAF of Abs to S100 suppressed generation of action potential in a dose-dependent manner and increased the maximal speed of its growth via changing the volt-ampere characteristics of the incoming current channels	[49]
	Involvement of GABAA-ergic system in the realization of RAF of Abs to S100 effects	Adult outbred male albino rats	Bicuculline and picrotoxin (GABAA-receptors antagonists) decreased the anxiolytic effect of RAF of Abs to S100 in Vogel conflict test by 1.8- and 1.6-times, respectively	[37]
	Involvement of GABAB-ergic system in the realization of RAF of Abs to S100 effects	Adult outbred male rats	Baclofen (GABAB-receptors agonist) decreased the anxiolytic effect of RAF of Abs to S100 in Vogel conflict test by 2.2-times, whereas phaclofen (GABAB-receptors antagonist) increased it by 1.4-times; both baclofen and phaclofen decreased antidepressive effect of RAF of Abs to S100 in Nomura's forced swimming test by 1.5- and 1.7-times, respectively	[50]
		CHO cells expressing human GABA receptors	RAF of Abs to S100 exerted antagonism at GABAB1A/B2-receptors inhibiting agonist-induced responses by 30.2% and also inhibited specific binding of ([3,4-3H]-cyclohexylmethyl)phosphinic acid ([3H]-CGP54626) to GABAB1A/B2-receptors by 25.8%	[51]
	Involvement of serotonergic system in the realization of RAF of Abs to S100 effects	Adult outbred male rats	Ketanserin (5-HT2 receptors antagonist) decreased both the anxiolytic effect of RAF of Abs to S100 in Vogel conflict test and antidepressive effect of RAF of Abs to S100 in Nomura forced swimming test by 1.9- and 2-times, respectively	[36]

Drug	Studied activity	Test system	Results	Ref
		CHO and CHO-K1 cells	RAF of Abs to S100 increased specific radioligands binding to 5-HT1F-, 5-HT2B-, 5-HT2Cedied- and 5-HT3-receptors by 142.0%, 131.9%, 149.3% and 120.7%, respectively; also RAF of Abs to S100 exerted antagonist effect at 5-HT1B-receptors inhibiting their functional activity by 23.2% and agonist effect at 5-HT1A-receptors enhancing their functional activity by 28.0%	[51]
	Involvement of dopaminergic system in the realization of RAF of Abs to S100 effects	CHO and CHO-K1 cells	RAF of Abs to S100 increased specific radioligand binding to D3-receptors by 126.3% and exerted antagonism at D3-receptors inhibiting their functional activity by 32.8%	[51]
	Involvement of glutamatergic system in the realization of RAF of Abs to S100 effects	Rat brain cortex neuronal cells	RAF of Abs to S100 decreased specific radioligand binding to NMDA-receptors by 39.1%	[52]
	Involvement of σ1-receptor in the realization of RAF of Abs to S100 effects	Human leukemic T-cell line (Jurkat); human breast cancer cell line (MCF-7)	RAF of Abs to S100 decreased specific radioligand binding to native and recombinant human σ1-receptors by 75.3% and 40.3%, respectively	[51]
RAF of Abs to eNOS	Effects on vascular endothelial function	Cavernous bodies of adult Wistar male rats	RAF of Abs to eNOS increased eNOS activity, the content of NO derivates and the content of cGMP by 2.4-, 1.3- and 4-times, respectively	[53]
Divaza	Involvement of σ1-receptor in the realization of Divaza effects	Segments of vas deferens of male albino Dunkin-Hartley guinea-pigs	Divaza increased the amplitude of tissue contraction induced by standard agonist by 2-times	[54]

Notes: * *- adapted from [73], CPAR - conditioned passive avoidance reflex, EPM - elevated plus maze, LPT - long-term potentiation.*

Appendix 1.
Experimental studies of pharmacological activity and mechanisms of action of RAF of Abs to S100, RAF of Abs to eNOS and combination drug Divaza.

Author details

Olga Vladimirovna Vorob'eva[1], Victoria Vyacheslavovna Fateeva[1,2*],
Ksenia Vladimirovna Nikulina[1,2], Kristina Konstantinovna Khacheva[2,3],
Gulnara Rinatovna Khakimova[2] and Oleg Ilyich Epstein[2,4]

1 Federal State Autonomous Educational Institution of Higher Education, I.M. Sechenov First Moscow State Medical University of the Ministry of Health of the Russian Federation (Sechenov University), Moscow, Russia

2 OOO «NPF «MATERIA MEDICA HOLDING»», Moscow, Russia

3 Research Center of Neurology, Moscow, Russia

4 The Institute of General Pathology and Pathophysiology, Moscow, Russia

*Address all correspondence to: v.v.fateeva@mail.ru

IntechOpen

© 2019 The Author(s). Licensee IntechOpen. This chapter is distributed under the terms of the Creative Commons Attribution License (http://creativecommons.org/licenses/by/3.0), which permits unrestricted use, distribution, and reproduction in any medium, provided the original work is properly cited.

References

[1] World Health Organization, Depression, Fact Sheet, 2017 (updated May 2019. http://www.who.int/mediacentre/factsheets/fs369/en/)

[2] Sassarini DJ. Depression in midlife women. Maturitas. 2016;**94**:149-154

[3] Dowlati Y, Herrmann N, Swardfager W, Liu H, Sham L, Reim EK, et al. A meta-analysis of cytokines in major depression. Biological Psychiatry. 2010;**67**(5):446-457

[4] Yu RH, Ho SC, Lam CW, Woo JL, Ho SS. Psychological factors and subclinical atherosclerosis in postmenopausal Chinese women in Hong Kong. Maturitas. 2010;**67**(2):186-191

[5] Chrysohoou C et al. The link between depression and atherosclerosis through the pathways of inflammation and endothelium dysfunction. Maturitas. 2018;**109**:1-5

[6] Nicholson A, Kuper H, Hemingway H. Depression as an aetiologic and prognostic factor in coronary heart disease: A meta-analysis of 6362 events among 146, 538 participants in 54 observational studies. European Heart Journal. 2006;**27**(23):2763-2774

[7] Kyrou I, Kollia N, Panagiotakos D, Georgousopoulou E, Chrysohoou C, Tsigos C, et al. Association of depression and anxiety status with 10-year cardiovascular disease incidence among apparently healthy Greek adults: The ATTICA study. European Journal of Preventive Cardiology. 2017;**24**(2):145-152

[8] Notara V, Panagiotakos DB, Papataxiarchis E, Verdi M, Michalopoulou M, Tsompanaki E, et al. Depression and marital status determine the 10-year (2004–2014) prognosis in patients with acute coronary syndrome: The GREECS study. Psychology & Health. 2015;**30**(9):1116-1127

[9] Antonogeorgos G, Panagiotakos DB, Pitsavos C, Papageorgiou C, Chrysohoou C, Papadimitriou GN, et al. Understanding the role of depression and anxiety on cardiovascular disease risk, using structural equation modeling; the mediating effect of the Mediterranean diet and physical activity: The ATTICA study. Annals of Epidemiology. 2012;**22**(9):630-637

[10] Mamplekou E, Bountziouka V, Psaltopoulou T, Zeimbekis A, Tsakoundakis N, Papaerakleous N, et al. Urban environment, physical inactivity and unhealthy dietary habits correlate to depression among elderly living in eastern Mediterranean islands: The MEDIS (MEDiterranean ISlands elderly) study. The Journal of Nutrition, Health & Aging. 2010;**14**(6):449-455

[11] Georgousopoulou EN, Kastorini CM, Milionis HJ, Ntziou E, Kostapanos MS, Nikolaou V, et al. Association between mediterranean diet and non-fatal cardiovascular events, in the context of anxiety and depression disorders: A case/case-control study. The Hellenic Journal of Cardiology. 2014;**55**(1):24-31

[12] Kollia N, Panagiotakos DB, Georgousopoulou E, Chrysohoou C, Yannakoulia M, Stefanadis C, et al. Exploring the path between depression, anxiety and 10-year cardiovascular disease incidence, among apparently healthy Greek middle-aged adults: The ATTICA study. Maturitas. 2017;**106**:73-79

[13] Panagiotakos DB, Pitsavos C, Chrysohoou C, Tsetsekou E,

Papageorgiou C, Christodoulou G, et al. ATTICA study, inflammation, coagulation, and depressive symptomatology in cardiovascular disease-free people; the ATTICA study. European Heart Journal. 2004;**25**(6): 492-499

[14] Vogelzangs N, Duivis HE, Beekman AT, Kluft C, Neuteboom J, Hoogendijk W. Association of depressive disorders, depression characteristics and antidepressant medication with inflammation. Translational Psychiatry. 2012;**2**:e79

[15] Smith RS. The macrophage theory of depression. Medical Hypotheses. 1991; **35**(4):298-306

[16] Kiecolt-Glaser JK, Derry HM, Fagundes CP. Inflammation: Depression fans the flames and feasts on the heat. American Journal of Psychiatry. 2015; **172**(11):1075-1091

[17] Jokela M, Virtanen M, Batty GD, Kivimaki M. Inflammation and specific symptoms of depression. JAMA Psychiatry. 2016;**73**(1):87-88

[18] Kopschina Feltes P, Doorduin J, Klein HC, Juárez-Orozco LE, Dierckx RA, Moriguchi-Jeckel CM. Anti-inflammatory treatment for major depressive disorder: Implications for patients with an elevated immune profile and non-responders to standard antidepressant therapy. Journal of Psychopharmacology. 2017;**31**(9): 1149-1165

[19] Felger JC. The role of dopamine in inflammation-associated depression: Mechanisms and therapeutic implications. Current Topics in Behavioral Neurosciences. 2017;**31**: 199-219

[20] Galecki P, Mossakowska-Wójcik J, Talarowska M. The anti-inflammatory mechanism of antidepressants-SSRIs, SNRIs. Progress in Neuro-Psychopharmacology and Biological Psychiatry. 2018;**80**(Pt C):291-294

[21] Pavlov VA, Tracey KJ. Controlling inflammation: The cholinergic anti-inflammatory pathway. Biochemical Society Transactions. 2006;**34**(Pt 6): 1037-1040

[22] Kong J, Fang J, Park J, Li S, Rong P. Treating depression with transcutaneous auricular vagus nerve stimulation: State of the art and future perspectives. Frontiers in Psychiatry. 2018;**9**:20

[23] Pajer K, Hoffman R, Gardner W, Chang CN, Boley D, Wang W. Endothelial dysfunction and negative emotions in adolescent girls. International Journal of Adolescent Medicine and Health. 2016;**28**(2): 141-148

[24] Munk PS, Isaksen K, Brønnick K, Kurz MW, Butt N, Larsen AI. Symptoms of anxiety and depression after percutaneous coronary intervention are associated with decreased heart rate variability, impaired endothelial function and increased inflammation. International Journal of Cardiology. 2012;**158**(1): 173-176

[25] Stillman AN, Moser DJ, Fiedorowicz J, Robinson HM, Haynes WG. Association of anxiety with resistance vessel dysfunction in human atherosclerosis. Psychosomatic Medicine. 2013;**75**(6):537-544

[26] Cepeda MS, Stang P, Makadia R. Depression is associated with high levels of C-reactive protein and low levels of fractional exhaled nitric oxide: Results from the 2007–2012 national health and nutrition examination surveys. The Journal of Clinical Psychiatry. 2016; **77**(12):1666-1671

[27] Sardinha A, Nardi AE. The role of anxiety in metabolic syndrome. Expert

Review of Endocrinology and Metabolism. 2012;**7**(1):63-71

[28] Miller GE, Freedland KE, Carney RM, Stetler CA, Banks WA. Pathways linking depression, adiposity, and inflammatory markers in healthy young adults. Brain, Behavior, and Immunity. 2003;**17**(4):276-285

[29] Liu C-H, Zhang G-Z, Li B, Li M, Woelfer M, Walter M, et al. Role of inflammation in depression relapse. Journal of Neuroinflammation. 2019; **16**:90

[30] Epstein O. The spatial homeostasis hypothesis. Symmetry. 2018;**10**(4):103. DOI: 10.3390/sym10040103

[31] Karpova GV, Fomina TI, Vetoshkina TV, Dubskaya TYU, Voronova OL, Timina EA. Doklinicheskie issledovaniya obshchetoksicheskih svojstv preparatov sverhmalyh doz antitel k endogennym regulyatoram. Byulleten eksperimental noj biologii i mediciny. 2009;**148**(8): 184-186

[32] Zhavbert ES, Surkova EI, YAkovleva NN, Dugina YL, Epshtejn OI, Bugaeva LI. Doklinicheskoe toksikologicheskoe issledovanie reliz-aktivnyh preparatov kak sposob prognozirovaniya ih farmakologicheskoj aktivnosti i pobochnyh effektov. Byulleten eksperimental noj biologii i mediciny. 2016;**161**(2):212-217

[33] Pankova TM, Starostina MV, Shtark MB, Epshtejn OI. Nejroprotektornos dejstvie sverhmalyh doz antitel k belku S100 v kul'ture nejroblastomy pri deprivacii kisloroda i glyukozy. Byulleten eksperimental noj biologii i mediciny. 2007;**144**(9): 260-263

[34] Voronina TA, Sergeeva SA, Martyushev-Poklad AV, Dugina JL, Epstein OI. Antibodies to S-100 protein in anxiety-depressive disorders in experimental and clinical conditions. Animal Models in Biological Psychiatry. 2006;**6**:137-152

[35] Epshtejn OI, Molodavkin GM, Voronina TA, Sergeeva SA. Antidepressivnye svojstva proprotena i amitriptilina: sravnitel'noe eksperimental'noe issledovanie. Byulleten eksperimental noj biologii i mediciny. 2003;**145**(6):34-36

[36] Hejfec IA, Dugina YL, Voronina TA, Molodavkin GM, Martyushev-Poklad AV, Sergeeva SA. Uchastie serotoninergicheskoj sistemy v mekhanizme dejstviya antitel k belku S-100 v sverhmalyh dozah. Byulleten eksperimental noj biologii i mediciny. 2007;**143**(5):535-537

[37] Voronina TA, Molodavkin GM, Sergeeva SA, Epshtejn OI. GAMK-ergicheskaya sistema v realizacii anksioliticheskogo dejstviya «Proprotena»: eksperimental'noe issledovanie. Byulleten eksperimental noj biologii i mediciny 2003;**1**:37-39

[38] Hejfec IA, Voronina TA, Dugina YL, Molodavkin GM, Sergeeva SA. Zavisimost' anksioliticheskoj aktivnosti ot uslovij provedeniya testa konfliktnoj situacii po VOGEL na primere tenotena i diazepama. Byulleten eksperimental noj biologii i mediciny. 2011;**3**:311-314

[39] Voronina TA, Molodavkin GM, Sergeeva SA, Epshtejn OI. Anksioliticheskij effekt "Proprotena" v usloviyah nakazuemogo i nenakazuemogo povedeniya. Byulleten eksperimental noj biologii i mediciny. 2003;**1**:31-33

[40] Belous AS, Arustamova AA, Pokrovskij MV, Korokin MV, Gudyrev OS, Belous VS. Ocenka korrekcii L-Name i gipoestrogen-inducirovannoj endotelial'noj disfunkcii preparatom impaza. Nauchnye vedomosti Belgorodskogo

gosudarstvennogo universiteta Seriya: Medicina Farmaciya. 2011;**4**(99): 116-120

[41] Arustamova AA. Protivoishemicheskoe i endotelioprotektivnoe dejstvie potencirovannyh antitel k endotelial'nomu faktoru rosta sosudov. Belgorod: Avtoref. Diss. ... kand. Biol. Nauk; 2011. p. 21

[42] Voronina TA, Molodavkin GM, ZHavbert ES, Tarasov SA, Hejfec IA, Dugina YUL. Anksioliticheskie i antidepressivnye svojstva preparata «Impaza». Byullyuten eksperimental noj biologii i mediciny. 2008;**145**(6): 682-684

[43] Markel' AL, ZHavbert ES, Tarasov SA, Dugina YUL, Kachanova MV, Sergeeva SA. Vliyanie impazy na serdechno-sosudistuyu sistemu. Byullyuten eksperimental noj biologii i mediciny. 2009;**148**(8):158-159

[44] Chernysheva GA, Aliev OI, Smol'yakova VI, Plotnikov MB, Martyushev AV, Sergeeva SA. Vliyanie impazy i sildenafila na serdechno-sosudistuyu sistemu i gipotenzivnyj effekt nitroglicerina. In: X Rossijskij nacional'nyj kongress «CHelovek i Lekarstvo». Moscow; S. P. Botkin National society for advanced training of doctor; 2003. p. 683

[45] Zhavbert ES, Gur'yanova NN, Surkova EI, Dugina YL, Kachaeva EV, Epshtejn OI. Issledovanie vliyaniya Divazy na processy perekisnogo okisleniya lipidov. In: XXI Rossijskij nacional'nyj kongress «CHelovek i lekarstvo». Moscow; S. P. Botkin National society for advanced training of doctor; 2014. p. 242

[46] Ganina KK, Dugina YUL, ZHavbert ES, Ertuzun IA, Epshtejn OI, Muhin VN. Antiamnesticheskoe dejstvie divazy i ee komponentov na modeli β-amiloidnoj amnezii u krys. ZHurnal nevrologii i psihiatrii. 2016;**9**:69-74

[47] Yakovleva NN, Voronina TA, Suslov NI, Ertuzun IA, Molodavkin GM, Poseva VI. Anksioliticheskoe i antidepressivnoe dejstvie preparatov «Divaza» i «Brizantin». Byulleten eksperimental noj biologii i mediciny. 2015;**159**(6):727-730

[48] Epshtejn OI, Beregovoj NA, Sorokina NS, Starostina MV, Shtark MB. Vliyanie razlichnyh razvedenij potencirovannyh antitel k mozgospecificheskomu belku S-100 na dinamiku posttetanicheskoj potenciacii v perezhivayushchih srezah gippokampa. Byulleten eksperimental noj biologii i mediciny. 2002;**127**(3): 317-320

[49] Epshtejn OI, Gajnutdinov HL, Shtark MB. Vliyanie gomeopaticheskih doz antitel k antigenu S-100 na elektricheskie harakteristiki nejronal'nyh membran. Byulleten eksperimental noj biologii i mediciny. 2002;**127**(4):48-51

[50] Hejfec IA, Molodavkin GM, Voronina TA, Dugina YUL, Sergeeva SA, Epshtejn OI. Uchastie GAMK-V-sistemy v mekhanizme dejstviya antitel k belku S-100 v sverhmalyh dozah. Byulleten eksperimental noj biologii i mediciny. 2008;**145**(5):552-554

[51] Gorbunov EA, Ertuzun IA, Kachaeva EV, Tarasov SA, Epstein OI. In vitro screening of major neurotransmitter systems possibly involved in the mechanism of action of antibodies to S100 protein in released-active form. Neuropsychiatric Disease and Treatment. 2015;**11**:2837-2846

[52] Ertuzun IA. Mekhanizmy anksioliticheskogo i antidepressatnogo dejstviya Tenotena (eksperimental'noe issledovanie). Avtoreferat dissertacii na

soiskanie uchenoj stepeni kandidata medicinskih nauk2012. p. 23

[53] Martyushev-Poklad AV, Epshtejn OI. Impaza – novoe lekarstvennoe sredstvo dlya lecheniya erektil'noj disfunkcii. Poliklinika. 2003; **1**(1):8-9

[54] Kardash EV, Gorbunov EA, Tarasov AV, YAkovleva NN, Tarasov SA. Vliyanie divazy na sigma receptory, reguliruyushchie osnovnye nejromediatornye sistemy. In: XXIII Rossijskij nacional'nyj kongress «CHelovek i lekarstvo». Moscow; S. P. Botkin National society for advanced training of doctor; 2016. p. 181

[55] Lanquillon S, Krieg JC, Bening-Abu-Shach U, Vedder H. Cytokine production and treatment response in major depressive disorder. Neuropsychopharmacology. 2000;**22**: 370-379

[56] Maes M. Cytokines in major depression. Biological Psychiatry. 1994; **36**:498-499

[57] Jokela M, Virtanen M, Batty GD, Kivimaki M. Inflammation and specific symptoms of depression. JAMA Psychiatry. 2016;**73**(1):87-88

[58] Musselman DL, Miller AH, Porter MR. Higher than normal plasma interleukin-6 concentrations in cancer patients with depression: Preliminary findings. The American Journal of Psychiatry. 2001;**158**:1252-1257

[59] Leonard BE. Brain cytokines and the psychopathology of depression. Antidepressants. 2001;**8**:109-120

[60] Frasure-Smith N, Lespérance F, Irwin MR. Depression, C-reactive protein and two-year major adverse cardiac events in men after acute coronary syndromes. Biological Psychiatry. 2007;**62**(4):302-308

[61] Copeland WE, Shanahan L, Worthman C. Cumulative depression episodes predict later C-reactive protein levels: A prospective analysis. Biological Psychiatry. 2012;**71**(1):15-21

[62] Wium-Andersen MK, Ørsted DD, Nielsen SF. Elevated C-reactive protein levels, psychological distress, and depression in 73 131 individuals. JAMA Psychiatry. 2013;**70**(2):176-184

[63] Howren MB, Lamkin DM, Suls J. Associations of depression with C-reactive protein, IL-a, and IL-6: A meta-analysis. Psychosomatic Medicine. 2009;**71**:171-186

[64] Dowlati Y, Herrmann N, Swardfager W. A meta-analysis of cytokines in major depression. Biological Psychiatry. 2010;**67**:446-457

[65] Frasure-Smith N, Lespérance F, Irwin MR. Depression, C-reactive protein and two-year major adverse cardiac events in men after acute coronary syndromes. Biological Psychiatry. 2007;**62**(4):302-308

[66] Hamer M, Batty GD, Marmot MG. Anti-depressant medication use and C-reactive protein: Results from two population-based studies. Brain, Behavior, and Immunity. 2011;**25**(1): 168-173

[67] Setiawan E, Wilson AA, Mizrahi R, Rusjan PM, Miler L, Rajkowska G, et al. Role of translocator protein density, a marker of neuroinflammation, in the brain during major depressive episodes. JAMA Psychiatry. 2015;**72**(3):268-275. DOI: 10.1001/jamapsychiatry.2014.2427

[68] Davignon J, Ganz P. Role of endothelial dysfunction in atherosclerosis. Circulation. 2004; **109**(23):III27-III32

[69] Verma S, Buchanan MR, Anderson TJ. Endothelial function

testing as a biomarker of vascular disease. Circulation. 2003;**108**: 2054-2059

[70] Rybakowski JK, Wykretowicz A, Heymann-Szlachcinska A, Wysocki H. Impairment of endothelial function in unipolar and bipolar depression. Biological Psychiatry. 2006;**60**(8): 889-891

[71] Broadley AJ, Korszun A, Jones CJ, Frenneaux MP. Arterial endothelial function is impaired in treated depression. Heart. 2002;**88**:521-523

[72] Do DP, Dowd JB, Ranjit N. Hopelessness, depression, and early markers of endothelial dysfunction in U. S. adults. Psychosomatic Medicine. 2010;**72**:613-619

[73] Parfenov VA, Ostroumova OD, Ostroumova TM, Kochetkov AI, Fateeva VV, Khacheva KK, et al. Vascular cognitive impairment: Pathophysiological mechanisms, insights into structural basis, and perspectives in specific treatments. Neuropsychiatric Disease and Treatment. 2019;**15**:1381-1402

Chapter 7

The Role of Vasoregulatory Markers in the Formation of Microcirculatory Changes in Premature Babies with Hypoxic: Ischemic Encephalopathy

Saadat Huseynova, Jamila Gurbanova, Afat Hasanova, Samaya Alizada and Nushaba Panakhova

Abstract

Endothelial function plays an important role in the extrauterine adaptation of newborn infants. Endothelium produces different biologically active mediators, which play the central role in physiological and pathological processes and also in the extrauterine adaptation of newborn infants. The imbalance between vasoconstrictive and vasodilatation factors results in impaired cardiovascular adaptation and microcirculation and also brain injury. Microcirculatory disturbances are observed very often in preterm babies, who have a serious risk for perinatal brain injury and further neurodevelopment disabilities. Present chapter presents the pathogenetic role of vascular tone regulators of endothelial genesis in the formation of microcirculatory changes in preterm babies with a high risk of perinatal hypoxic encephalopathy.

Keywords: microcirculation, hypoxic encephalopathy, preterm infants, endothelial function, nitric oxide

1. Introduction

The largest research object of modern perinatology and neonatology is preterm and growth retarded children. Despite the rapid development of perinatal care and the early prevention of many pathologies, worldwide perinatal morbidity and mortality remain high [1–4].

The results of the scientific researches prove that perinatal pathologies play a leading role in the formation of illness, death, disability, social and biological disarray, and different types of neurodevelopmental disorders [5–10]. It is known that birth is a complicated biological process regulated by numerous signal molecules and biologically active substances. The fetal inflammatory response plays a major role in the pathogenesis of premature birth [11]. In addition to prematurity, the hypoxic-ischemic changes in feto-placental system can result in different perinatal pathologies, such as acute intraventricular bleeding, periventricular leukomalation, necrotic enterocolitis, bronchial lung dysplasia, myocardial dysfunction, sepsis, etc. [12–15].

IntechOpen

Uteroplacental ischemia and circulatory changes in maternal-fetal system are the main chain in formation intrauterine hypoxia and different perinatal pathologies [16–18]. Previous investigations confirmed the significant role of endothelial function in the formation of different pregnancy pathologies and birth defects [19–22]. The pathogenetic mechanisms of the formation of endothelial dysfunction during uteroplacental ischemia have not yet been investigated. Present chapter explores the role of vascular tone regulators of endothelial genesis in formation of microcirculatory and ischemic changes in preterm infants.

2. The pathophysiology of brain injury in hypoxic: ischemic encephalopathy (HIE)

Adaptation of the child to the extrauterine life significantly depends on the morpho-functional maturity of the organism, and it is more intense and more complicated in preterm babies than mature children [23–25].

The progress of all complicated pathophysiologic processes occurring in the newborn after birth significantly depends on cardiorespiratory adaptation [23, 24]. The changes in the cardiovascular and respiratory functions in the body related to the primarily changes in the microcirculation [25]. Microcirculatory changes are not only clinical symptoms of various pathologies of perinatal period but also one of the major factors that aggravate their course [26, 27].

HIE is one of the most serious birth complications accompanying with microcirculatory changes of different severity [28]. The pathogenesis of vascular changes in preterm infants is quite complicated and involves series of biochemical and molecular reactions (**Figure 1**). Persistent membrane depolarization results in excessive presynaptic glutamate release which follows with a series of cellular changes. The activation of NMDA receptors stimulates profound Ca^{2+} influx, which mediates cascades to cell death. *Primary energy failure* associated with the depletion of oxygen prevents oxidative phosphorylation, and the disrupting Na-K pump activity is

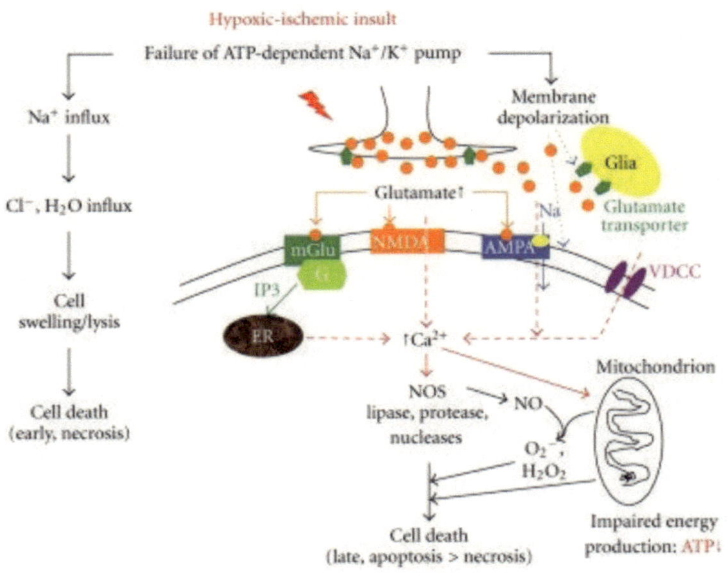

Figure 1.
The pathogenesis of hypoxic-ischemic encephalopathy [33].

followed by anaerobic metabolism with accumulation of lactic acid. With the restoration of blood flow, there is a brief period of normalization of cerebral metabolism called a latent period. The reperfusion is necessary for the recovery and stopping of processes leading to necrotic neuronal injury during the primary phase of injury. However when the brain has not recovered from the initial injury, the reperfusion can simultaneously cause additional (delayed) injury, and mitochondrial dysfunction continues. When cerebral ischemia is more acute and prolonged, especially in the result of accompanying pathological processes (infection, hereditary factors, environmental and other damaging factors), primary injury is followed by secondary injury, which is often characterized by subsequent resulting in more serious neurological and somatic disintegration in development [29]. Secondary injury is often associated with edema of the brain cells. Compensatory restoration of energy reactions is followed by the intracellular edema and by more complex inflammatory response cascade with the presence of free oxygen radicals [30].

Increased amount of free radicals and nitric oxide (NO), increased synthesis of nitric oxide synthase (NOS), activated intercellular adhesion, and apoptosis are the tightly connected chains of this pathological process (10–13). However it is confirmed that endothelial NOS (eNOS) genesis plays very important role in maintaining pulmonary blood flow and preventing pulmonary hypertension. Some experimental studies suggested that *inhibiting NOS could prevent further brain injury* [31]. Selective inhibition of NO of neuronal genesis is more promising in the direction of pathogenetic treatment of HIE in newborn infants [31–33].

The severity of inflammatory processes is correlated with the activation of different mediators, especially cytokines and adhesion molecules. These molecules cause to the migration of leukocytes to the inflammation center and compact adhesion of migrated leukocytes to vascular endothelium [34, 35]. The main stimulus factor for the synthesis of inflammatory mediators is the activation of endothelial cells of the fetus. Thus, endothelial dysfunction is the main factor that stimulates intracellular and vascular adhesion and leads to the activation of fetal leukocytes [36, 37].

There is much to be investigated how the inflammatory response to hypoxia is regulated and the complete role of different mediators as well as vasoregulatory, anti-inflammatory, and apoptosis molecules under physiological and pathological conditions is unknown. The goal of this chapter is to present the results of recent investigations about the role of vasoregulatory markers in the formation of microcirculatory disorders in hypoxic-ischemic encephalopathy of preterm infants.

3. Endothelial dysfunction and microcirculatory disorders in HIE of preterm infants

Several clinical and experimental studies confirmed the role of endothelial dysfunction in the pathogenesis of hypoxic-ischemic brain injury. The prospective clinical trial of Azerbaijan Medical University Neonatology group (ACTRN12612000342819) determined that the eNOS activity is declined in the background of increased NO concentrations depending on the severity of HIE [38].

The aim of the same study was also to study of the peripheral blood concentrations of vasoregulatory mediators of endothelial genesis in the pathogenesis of microcirculatory changes in newborn children with the birth asphyxia. It investigated 240 preterm infants with a high risk of HIE during early neonatal period. The main groups of children were classified into four groups depending on the degree of the microcirculation changes. The first group included preterm infants without microcirculatory changes of the body. The children with mild-degree microcirculatory disorders (continued less than 1 day and self-regenerating peroral and acrocyanosis, capillary refilling time duration less than 3 s) were included in the second group. The third group consisted of children with moderate microcirculatory disorders (such as peroral and acrocyanosis, marbling of the skin, capillary refilling time up to 7 s and continuing from 1 day up to 3 days). The fourth group consisted of children with severe microcirculatory disorders (acute peroral and acrocyanosis, marbling of the skin continuing more than 3 days, capillary refilling time with the duration of more than 7 s and continuing more than 3 days). The parameters were compared with the data of 2 control groups, which consisted of infants without perinatal and neonatal pathologies: 22 healthy preterm infants were included in control 1 and 30 healthy term infants in control 2.

Depending on the magnitude of the microcirculatory defects, the levels of vasoregulatory markers included in the study is shown in **Table 1**. The statistically significant reduction in eNOS activity in the first few days of life is noticeable, depending on the degree of severity of the microcirculatory disturbances. However toward the end of the early neonatal period in mild and moderate group children, eNOS concentrations significantly increased compared with children with severe microcirculatory changes and control groups.

As shown in **Table 1**, during severe microcirculation defects, NO synthesis of vascular endothelium remains at very low levels. In contrast, NO levels in the early days of the neonatal period were noted to significantly increase in infants with severe microcirculatory disturbances, and in the dynamics of the neonatal period, regardless of the microcirculatory changes severity, it is observed the increase of NO concentrations. At the same time, vasoconstrictor endothelin-1 levels rise during mild and moderate grades of microcirculation changes, while in infants with severe changes, it is reduced. This also proves once again that severe microcirculation disturbances lead to a violation of blood supply both in peripheral and vital organs during acute brain damage. We suggest that the lack of adequate levels of endothelin-1 synthesis, which is vasoconstrictor mediator of vascular endothelium in addition to decreased endothelial NOS activity, becomes one of the main points in the pathophysiology of HIE in preterm infants.

The follow-up results of these children included in this study identified significant relationships between peripheral endothelial vasoregulatory markers in the perinatal period and the formation of developmental disorders at an early age [39]. It was found that, in the presence of high concentrations of NO, early eNOS activity was insufficient in infants with moderate-to-severe neurodevelopmental disorders compared to neonates with mild neurologic changes or without evidence of neurological impairment (**Table 2**). These findings suggest that depressed eNOS activity and increased non-endothelial NO synthesis play also important roles in the formation of developmental impairments.

It is known that there is a disturbance of vasoregulation in the pathogenesis of various pathologies of the HIE and prenatal period [40–42]. Depending on the complexity of the pathological process and the degree of morphologic and functional immaturity of the body, hypoxic-ischemic lesions can lead to generalized system damage from mild to generalized severe dysfunctions and changes [43–47]. Acceleration of blood supply to vital organs during HIE is accompanied by peripheral vasospasm. However, the depletion of vascular tone's regulating mechanisms during the severe and long-lasting processes leads to the tissue hypoxia and acidosis [40–47]. This often leads to changes in vital organs, especially in brain tissue whose results are with changes that cannot be restored.

It is considered that statistically significant increase of NO levels in peripheral blood circulation during severe hypoxic changes is due to the exhaustion of endothelial NOS sources and the activation of non-endothelial NO synthesis sources.

DOI: http://dx.doi.org/10.5772/intechopen.89910

		N	Mean	SE	p<0.05	95% CI		Min	Max
eNOS 1-3rd day	none	20	10,28	1,24	1,2,3,#,^	7,67	12,89	0,30	21,20
	mild	28	6,65	1,12	0,3,#^	4,35	8,96	0,00	22,20
	moderate	28	5,84	0,91	0,#,^	3,97	7,72	0,04	13,50
	severe	18	3,10	0,79	0	1,43	4,77	0,30	13,50
	control 1	22	2,11	0,09		1,92	2,31	1,50	2,90
	control 2	30	1,98	0,06		1,84	2,12	1,50	2,70
eNOS 7-10th day	none	4	2,56	0,86	^	0,18	5,31	0,06	4,00
	mild	17	5,29	1,49	#,^	2,11	8,46	0,06	24,40
	moderate	25	5,23	1,17		2,79	7,66	0,06	24,40
	severe	11	2,14	0,27		1,53	2,75	0,70	3,70
	control 1	22	1,80	0,06		1,67	1,93	1,30	2,30
	control 2	30	1,33	0,04		1,23	1,43	0,90	2,20
NO 1-3rd day	none	24	29,47	2,24	2,3	24,82	34,12	16,00	53,40
	mild	49	36,41	1,05	2,3,^	34,29	38,52	16,00	49,20
	moderate	58	48,46	2,89	0,1,3,#,^	42,67	54,25	16,00	99,40
	severe	24	70,40	5,56	0,1,2,#,^	58,88	81,91	22,00	99,40
	control 1	22	27,95	0,66		26,57	29,33	23,00	35,00
	control 2	30	23,20	0,57		22,02	24,37	16,00	31,00
NO 7-10th day	none	20	35,73	3,49	^	28,42	43,03	0,80	50,80
	mild	50	27,05	2,28	2,^	22,46	31,63	0,80	50,80
	moderate	39	36,12	2,44	1,5,^	31,18	41,07	0,80	50,80
	severe	6	36,06	4,67	^	24,05	48,08	25,60	50,80
	control 1	22	25,36	0,77		23,74	26,97	16,00	32,00
	control 2	30	17,43	0,63		16,13	18,72	10,00	23,00
ET-1 1-3rd day	none	29	3,23	0,65	3,^	1,89	4,56	0,18	12,18
	mild	49	4,58	0,55	^	3,47	5,69	0,21	13,02
	moderate	58	4,21	0,57	1	3,06	5,36	0,21	13,02
	severe	29	2,32	0,49		1,32	3,33	0,63	13,02
	control 1	22	2,83	0,09		2,63	3,02	1,85	3,90
	control 2	30	1,85	0,05		1,73	1,97	1,10	2,40
ET-1 7-10th day	none	22	2,16	0,25		1,63	2,70	0,84	6,49
	mild	38	2,69	0,35		1,98	3,41	0,42	9,93
	moderate	45	3,43	0,38		2,66	4,21	1,05	9,93
	severe	24	2,90	0,32		2,23	3,57	0,42	7,36
	control 1	22	1,88	0,05		1,77	2,00	1,40	2,30
	control 2	30	1,53	0,07		1,37	1,68	0,90	2,30

Table 1.
The level of vasoregulatory indicators in microcirculatory disturbances in children with HIE risk (p<0,05 in comparison with children with 0-none of, 1- mild, 2-moderate, 3-severe microcirculatory changes, and with # - Control 1, ^ - Control 2 infants).

It is likely that in high endothelin-1 levels in children, mild and moderate changes are likely to compensate for an increase in peripheral vein tone and vital organs to maintain normal blood circulation. Reduced vasoconstrictor endothelium-1 levels in children with severe HIE symptoms are likely to be associated with decreased vascular tone and tissue hypoperfusion. In conclusion, the changes of capillary blood circulation in the result of endothelial dysfunction have the main role in the pathogenesis of hypoxic-ischemic inflammation in preterm infants.

Variables	Groups	N	Mean±SD	Minimum	Maximum	^{a}p value
eNOS, IU/ml, day1-3	1st group	8	4,65±1,67	2,50	7,60	$^{1\text{-}c}p<0,01$ $^{2\text{-}c}p<0,01$ $^{3\text{-}c}p<0,01$
	2nd group	20	7,71±3,55	3,20	15,00	
	3rd group	14	6,71±3,83	2,80	15,00	
	control group	20	2,09±0,63	0,70	3,20	
eNOS, IU/ml, day7-10	1st group	8	1,37±0,43	0,70	1,90	$^{1\text{-}2}p<0,05$ $^{1\text{-}3}p<0,05$ $^{2\text{-}3}p<0,05$ $^{2\text{-}c}p<0,05$ $^{3\text{-}c}p<0,01$
	2nd group	20	2,64±1,01	0,60	4,10	
	3rd group	14	7,57±3,27	3,90	14,00	
	control group	20	1,85±0,74	0,50	3,20	
NO, mmol/1, day1-3	1st group	8	47,50±17,51	23,00	72,00	$^{1\text{-}c}p<0,05$ $^{2\text{-}c}p<0,01$ $^{3\text{-}c}p<0,01$
	2nd group	20	45,60±12,39	23,00	72,00	
	3rd group	14	42,85±13,40	19,00	68,00	
	control group	20	25,00±7,72	16,00	41,00	
NO, mmol/1, day7-10	1st group	8	69,37±13,74	45,00	87,00	$^{1\text{-}3}p<0,05$ $^{2\text{-}3}p<0,05$ $^{1\text{-}c}p<0,01$ $^{2\text{-}c}p<0,05$ $^{3\text{-}c}p<0,01$
	2nd group	20	65,00±31,04	17,00	108,00	
	3rd group	14	43,57±14,71	19,00	68,00	
	control group	20	26,50±7,97	16,00	41,00	
Endothelin-1, pg/ml, day1-3	1st group	8	5,71±2,60	3,20	7,20	$^{2\text{-}3}p<0,05$ $^{1\text{-}c}p<0,01$ $^{2\text{-}c}p<0,01$ $^{3\text{-}c}p<0,01$
	2nd group	20	4,43±1,55	1,30	6,70	
	3rd group	14	5,70±1,73	3,20	7,90	
	control group	20	2,71±0,56	1,60	3,90	
Endothelin-1, pg/ml, day7-10	1st group	8	3,92±1,08	2,60	5,60	$^{1\text{-}2}p<0,05$ $^{2\text{-}3}p<0,05$ $^{1\text{-}c}p<0,01$ $^{2\text{-}c}p<0,01$ $^{3\text{-}c}p<0,01$
	2nd group	20	5,91±1,61	3,20	8,60	
	3rd group	14	4,02±0,93	2,60	5,60	
	control group	20	1,82±0,69	,90	3,90	

Table 2.
Blood concentrations of vasoregulatory markers in the early neonatal period by study groups. 1st group: HIE infants diagnosed with moderate-to-severe neurodevelopmental disorders or cerebral palsy; 2nd group: HIE infants with mild neurologic changes at an early age; 3rd group: HIE infants without evidence of neurological impairment in the post-neonatal period; control group: healthy preterm infants. $^{a}p<0,05$ is considered statistically significant between main groups (1-2, 1-3, 2-3), and between main and control groups (1-c, 2-c, 3-c).

Acknowledgements

The authors sincerely thank Science Development Foundation under the President of the Azerbaijan Republic for providing specific reagent kits (Grant SDF-2010-1(1)-40/28-M-2). We also thank the staff of the Clinical Biochemistry Laboratory of Azerbaijan Medical University for assistance with biomarker analysis.

Conflict of interest

The authors declare no conflict of interests.

Author details

Saadat Huseynova[1]*, Jamila Gurbanova[1], Afat Hasanova[1], Samaya Alizada[1] and Nushaba Panakhova[2]

1 Scientific Research Institute of Obstetrics and Gynecology, Baku, Azerbaijan

2 Pediatrics Department of Azerbaijan Medical University, Baku, Azerbaijan

*Address all correspondence to: sadi_0105@mail.ru

IntechOpen

© 2019 The Author(s). Licensee IntechOpen. This chapter is distributed under the terms of the Creative Commons Attribution License (http://creativecommons.org/licenses/by/3.0), which permits unrestricted use, distribution, and reproduction in any medium, provided the original work is properly cited.

References

[1] Lees C, Marlow N, Arabin B, Bilardo CM, Brezinka C, Derks JB, et al. Perinatal morbidity and mortality in early-onset fetal growth restriction: Cohort outcomes of the trial of randomized umbilical and fetal flow in Europe (TRUFFLE). Ultrasound in Obstetrics & Gynecology. 2013;**42**:400-408

[2] Unterscheider J, Daly S, Geary MP, Kennelly MM, McAuliffe FM, O'Donoghue K, et al. Optimizing the definition of intrauterine growth restriction: The multicenter prospective PORTO study. American Journal of Obstetrics and Gynecology. 2013;**208**(290):e1-e6

[3] Bader D, Kugelman A, Boyko V, Levitzki O, Lerner-Geva L, Riskin A, et al. Risk factors and estimation tool for death among extremely premature infants; a national study. Pediatrics. 2010;**125**:696-703

[4] Mooney SS, Lee RM, Tong S, Brownfoot FC. Expectant management of severe preterm preeclampsia: A comparison of maternal and fetal indications for delivery. Journal of Maternal-Fetal and Neonatal Medicine. 2016;**29**(23):1-6. DOI: 10.3109/14767058.2016.1147555

[5] Baschat AA. Neurodevelopment following fetal growth restriction and its relationship with antepartum parameters of placental dysfunction. Ultrasound in Obstetrics & Gynecology. 2011;**37**:501-514

[6] Morsing E, Asard M, Ley D, Stjernqvist K, Marsal K. Cognitive function following intrauterine growth restriction and very preterm birth. Pediatrics. 2011;**127**:874-882

[7] Morsing E, Gustafsson P, Brodzki J. Lung function in children born after fetal growth restriction and very preterm birth. Acta Paediatrica. 2012;**101**:48-54

[8] Markestad T, Kaaresen PI, Rønnestad A, et al. Early death, morbidity, and need of treatment among extremely premature infants. Pediatrics. 2005;**115**:1289

[9] Schmidt B, Roberts RS, Davis PG, et al. Prediction of late death or disability at age 5 years using a count of 3 neonatal morbidities in very low birth weight infants. The Journal of Pediatrics. 2015;**167**:982

[10] Goldenberg RL, Gravett MG, Iams J, Papageorghiou AT, Waller SA, Kramer M, et al. The preterm birth syndrome: Issues to consider in creating a classification system. American Journal of Obstetrics and Gynecology. 2012;**206**(2):113-118. DOI: 10.1016/j.ajog.2011.10.865

[11] Kadhim HJ, Duchateau J, Sebire G. Cytokines and brain injury: Invited review. Journal of Intensive Care Medicine. 2008;**23**:236-249

[12] Malaeb S, Dammann O. Fetal inflammatory response and brain injury in the preterm newborn. Journal of Child Neurology. 2009;**24**(9):1119-1126

[13] Dammann O, Leviton A. Inflammatory brain damage in preterm newborns--dry numbers, wet lab, and causal inferences. Early Human Development. 2004;**79**:1-15

[14] Sciaky-Tamir Y, Hershkovitz R, Mazor M, Shelef I, Erez O. The use of imaging technology in the assessment of the fetal inflammatory response syndrome—Imaging of the fetal thymus. Prenatal Diagnosis. 2015;**35**(5):413-419

[15] Viscardi RM. Perinatal inflammation and lung injury. Seminars

in Fetal & Neonatal Medicine. 2012 Feb;**17**(1):30-35

[16] Jeffrey S, Gilbert AJB, Gingery A, Chasson S. Circulating and utero-placental adaptations to chronic placental ischemia in the rat. Placenta. 2011;**33**(2):100-105

[17] Espinoza J. Uteroplacental ischemia in early- and late-onset pre-eclampsia: A role for the fetus? Ultrasound in Obstetrics & Gynecology. 2012;**40**(4):373-382

[18] Warrington J, Fan F, Murphy S, Roman R, Drummond H, Granger J, et al. Placental ischemia impairs cerebral blood flow autoregulation and increases blood-brain barrier permeability in pregnant rats. The FASEB Journal. 2015;**29**(1 Supplement):646.7

[19] Gilbert JS, Ryan MJ, LaMarca BB, Sedeek M, Murphy SR, Granger JP. Pathophysiology of hypertension during preeclampsia: Linking placental ischemia with endothelial dysfunction. American Journal of Physiology - Heart and Circulatory Physiology. 2008;**294**(2):H541-H550

[20] Granger JP, Alexander BT, Llinas MT, Bennett WA, Khalil RA. Pathophysiology of preeclampsia: Linking placental ischemia/hypoxia with microvascular dysfunction. Microcirculation. 2002;**9**(3):147-160

[21] Sanchez L, Prada CE, Riaño-Medina CE, Lopez M. Endothelial dysfunction and preeclampsia: Role of oxidative stress. Frontiers in Physiology. 2014;**5**:372

[22] Zárate A, Saucedo R, Valencia J, Manuel L, Hernández M. Early disturbed placental ischemia and hypoxia creates immune alteration and vascular disorder causing preeclampsia. Archives of Medical Research. 2014;**45**(7):519-524

[23] MacDonald JF, Xiong ZG, Jackson MF. Paradox of Ca2+ signaling, cell death and stroke. Trends in Neurosciences. 2006;**29**:75-81

[24] Executive summary. Neonatal encephalopathy and neurologic outcome, second edition. Report of the American College of Obstetricians and Gynecologists' task force on neonatal encephalopathy. The Obstetrician and Gynaecologist. 2014;**123**:896-901

[25] O'Sullivan M. Leukoaraiosis. Practical Neurology. 2008;**8**:26-38

[26] Wassink G, Gunn ER, Drury PP, et al. The mechanisms and treatment of asphyxial encephalopathy. Frontiers in Neuroscience. 2014;**8**:40

[27] Peters A, Schweiger U, Pellerin L, et al. The selfish brain: Competition for energy resources. Neuroscience and Biobehavioral Reviews. 2004;**28**:143-180

[28] Martinez-Biarge M, Diez-Sebastian J, Wusthoff CJ, Mercuri E, Cowan FM. Antepartum and intrapartum factors preceding neonatal hypoxic-ischemic encephalopathy. Pediatrics. 2013;**132**(4):e952-e959

[29] Northington FJ, Chavez-Valdez R, Martin LJ. Neuronal cell death in neonatal hypoxia-ischemia. Annals of Neurology. 2011 May;**69**(5):743-758

[30] Hagberg H, Mallard C, Ferriero DM, Vannucci SJ, Levison SW, Vexler ZS, et al. The role of inflammation in perinatal brain injury. Nature Reviews Neurology. 2015;**11**:192-208

[31] Pham H, Duy AP, Pansiot J, Bollen B, Gallego J, Charriaut-Marlangue C, et al. Impact of inhaled nitric oxide on white matter damage in growth-restricted neonatal rats. Pediatric Research. 2015;**77**(4):563-569

[32] Ji H, Tan S, Igarashi J, et al. Selective neuronal nitric oxide synthase inhibitors

and the prevention of cerebral palsy. Annals of Neurology;**65**(2):209-217

[33] Lai M-C, San-Nan YJ. Perinatal hypoxic-ischemic encephalopathy. Journal of Biomedicine and Biotechnology. 2011;**2011**:609813

[34] Buschmann K, Tschada R, Metzger MS, Braach N, Kuss N, Hudalla H, et al. RAGE controls leukocyte adhesion in preterm and term infants. BMC Immunology. 2014;**15**:53. DOI: 10.1186/s12865-014-0053-0

[35] Nussbaum C, Gloning A, Pruenster M, Frommhold D, Bierschenk S, Genzel-Boroviczény O, et al. Neutrophil and endothelial adhesive function during human fetal ontogeny. Journal of Leukocyte Biology. 2013;**93**(2):175-184

[36] D'Alquen D, Kramer BW, Seidenspinner S, Speer CP. Activation of umbilical cord endothelial cells and fetal inflammatory response in preterm infants with chorioamnionitis and funisitis. Pediatric Research. 2005;**57**(2):263-269

[37] Liao JK. Linking endothelial dysfunction with endothelial cell activation. The Journal of Clinical Investigation. 2013;**123**(2):540-541

[38] Huseynova S, Panakhova N, Orujova P, Hasanov S, Guliyev M, Orujov A. Elevated levels of serum sICAM-1 in asphyxiated low birth weight newborns. Scientific Reports. 2014;**4**:6850. DOI: 10.1038/srep06850

[39] Huseynova SA, Panakhova NF, Hasanov SS, Guliyev MR. Nitric oxide synthase and neurodevelopmental disorders. In: Nitric Oxide Synthase. IntechOpen; 2016. ISBN 978-953-51-5012

[40] Graham EM, Ruis KA, Hartman AL, Northington FJ, Fox HE. A systematic review of the role of intrapartum hypoxiaischemia in the causation of neonatal encephalopathy. American Journal of Obstetrics and Gynecology. 2008;**199**(6):587-595

[41] Azzopardi D. Clinical management of the baby with hypoxic ischaemic encephalopathy. Early Human Development. 2010;**86**(6):345-350

[42] Mercuri E, Guzzetta A, Haataja L, Cowan F, Rutherford M, Counsell S, et al. Neonatal neurological examination in infants with hypoxic ischaemic encephalopathy: Correlation with MRI findings. Neuropediatrics. 1999;**30**(2):83-89

[43] MacLennan AH, Thompson SC, Gecz J. Cerebral palsy: Causes, pathways, and the role of genetic variants. American Journal of Obstetrics and Gynecology. 2015;**213**:779-788

[44] Beken S, Aydın B, Dilli D, Erol S, Zenciroğlu A, Okumuş N. Can biochemical markers predict the severity of hypoxicischemic encephalopathy? The Turkish Journal of Pediatrics. 2014;**56**(1):62-68

[45] Northington FJ, Zelaya ME, O'Riordan DP, et al. Failure to complete apoptosis following neonatal hypoxia-ischemia manifests as "continuum" phenotype of cell death and occurs with multiple manifestations of mitochondrial dysfunction in rodent forebrain. Neuroscience. 2007;**149**(4):822-833

[46] Volpe JJ. Neurology of the Newborn. 5th ed. Saunders; 2008. 1120 p

[47] Vannucci RC, Towfighi J, Vannucci SJ. Secondary energy failure after cerebral hypoxia-ischemia in the immature rat. Journal of Cerebral Blood Flow and Metabolism. 2004;**24**(10):1090-1097

Chapter 8

Functional State of the Microvascular Bed of the Skin in Essential Arterial Hypertension Assessed by Laser Doppler Flowmetry with Amplitude-Frequency Wavelet Analysis of Blood Flow Oscillations

Andrey A. Fedorovich

Abstract

Modern diagnostic technologies provide access to data which until recently were not available to specialists. Laser Doppler flowmetry (LDF) with amplitude-frequency wavelet analysis of tissue perfusion oscillations in patients with essential arterial hypertension (AH) revealed two completely opposite functional states of resistive precapillary arterioles. One group of patients (48%) had a significant increase in the sympathetic vascular tone component and vasomotor microvascular endothelial dysfunction (precapillary resistance). The second group of patients (52%), on the contrary, exhibited a significant reduction in myogenic tone accompanied by raised arterial blood inflow and increased blood filling of the venular microcirculatory bloodstream (postcapillary resistance). Since the main target of most pharmaceuticals is the resistive vessels, LDF can be useful for the selection of personalized antihypertensive therapy that considers the functional state of resistive microvessels.

Keywords: vasomotion, arteriole tone, peripheral vascular resistance, arterial hypertension, laser Doppler flowmetry, wavelet analysis

1. Introduction

In the past two decades, the interest in microcirculatory blood flow in patients with arterial hypertension (AH) has been steadily increasing due to the significant role of microcirculatory disorders in the pathogenesis of this disease [1–3] and unsatisfactory treatment results, when despite achievement of the target blood pressure with antihypertensive therapy, residual cardiovascular risks remain substantial [4]. Laser Doppler flowmetry (LDF) is a modern noninvasive technique for examining the microcirculatory blood flow in humans. Even though the skin vessels are not subjected to baroreflex regulation, the accumulated data suggests that the microvascular bed of the skin may reflect the microcirculatory system state in other bodily organs and systems [5–8]. The results of functional LDF tests demonstrate a significant correlation between the left ventricular ejection fraction and the end-diastolic volume [9], flow-mediated vasodilation [10], and renal resistive index [11]. Discontinuation of antihypertensive therapy leads to a decrease in post-occlusive reactive hyperemia [12], which, in turn, correlates with the cardiovascular risk factors in the female population [13]. In 2011, the Peripheral Circulation Working Group of the European Society of Cardiology included LDF in the list of the recommended methods to study endothelial function [14].

Arterial hypertension is a hemodynamic disease with a blood pressure (BP) rise due to an increase in the cardiac output and/or peripheral vascular resistance [15]. In the classic work on normo-, hypo-, and hypertensive cats, Zweifach demonstrated that in the mesentery microvessels the greatest pressure gradient is recorded in arterioles <50 μm in diameter [16, 17]; at this level the Reynolds number is less than one and the viscous blood forces prevail over kinetic ones [18]. As in LDF resistive microvessels <50 μm in diameter are included in the probed volume [19], the interest in their functional state in AH is quite natural.

A fundamental feature of resistive arterioles is their high vasomotor activity. The arterioles are in constant motion, changing their tone and their lumen size, which manifests as vasomotions [20] with the respective changes in tissue perfusion. The vasomotion phenomenon is due to the ability of smooth muscle cells to spontaneously contract with an average frequency of 6 times per minute. It is the myogenic resistance at the capillary sphincter level that is the last blood flow control link before the exchange vessels, i.e. capillaries. Myogenic vasomotions are clearly conducted into the capillary bed of the human skin [21], and their amplitude is positively correlated with the number of functioning capillaries [22]. In arterioles, the basal tone and vasomotor activity of smooth muscle cells are modulated from the outer layer of the vessel wall by sympathetic nervous system 2–3 times per minute, and from the lumen by endothelial factors with a frequency of less than once per minute. The vascular tone-forming mechanisms (endothelial, neurogenic, and myogenic) act directly via the smooth muscle cells of microvessels and, as a result of periodic changes in blood flow resistance, generate the corresponding fluctuations in tissue perfusion [23, 24]. Due to the alternating contractions and relaxations of the smooth muscle in the arterioles and capillary sphincters, the arterial blood flowing into the capillaries is modulated to the optimal volume for transcapillary exchange [25–28]. During self-organization of microcirculation, all regulatory mechanisms interact with each other in positive and negative feedback loops that are aimed at maintaining tissue homeostasis. Thus, the tone-forming mechanisms function mainly at the resistive arteriole level, thereby determining not only the capillary hemodynamic parameters but also the peripheral vascular resistance (PVR).

Other microcirculation modulation mechanisms are passive in relation to the arteriolar smooth muscle cells, but they determine the blood filling volume of the microvascular bed (MVB) by changing the longitudinal pressure gradient caused by periodic changes in BP at the "inlet" (pulse BP) and pressure variation in the venules during the respiratory cycles at the "outlet" of MVB. The pulse oscillation amplitude reflects the condition of the inflow tracts (arterioles) and the arterial blood inflow into the MVB, and the amplitude of respiratory-associated blood flow oscillations characterizes the state of the capillary outflow tract, thereby reflecting the blood volume in the venular section of MVB [26, 29–31].

The main aim of this pilot study was to evaluate the functional state of resistive vessels of the skin in patients with essential arterial hypertension according by LDF with an amplitude-frequency wavelet analysis of microcirculation fluctuations.

2. Materials, methods and results

The skin perfusion was assessed in the supine position after a 15-min period of adaptation in a laboratory with a constant microclimate (+23 ± 1°C) in the morning (09:00–12:00 AM). The microcirculatory blood flow was recorded by a LAKK-02 single-channel laser analyzer of blood microcirculation in the visible red spectrum (wavelength of 630 nm) and a LAKK-TEST complex (OOO Research and Production Enterprise LAZMA, Russia) that allow evaluating the perfusion parameters in ~ 1.0–1.5 mm^3 of skin, at a constant temperature in the studied region of +32°C. The sensor was located on the outer surface of the right forearm, in the midline 3–4 cm proximal to the wrist joint. The perfusion was recorded for 6 min. After the microcirculation study, all subjects received 24-h ambulatory blood pressure monitoring (ABPM) on the left shoulder.

The initial LDF recording (**Figure 1A**) was subjected to amplitude-frequency wavelet analysis. The time-averaged amplitude of vasomotions was estimated by the maximum values (Amax) in the corresponding [32, 33] blood flow modulation frequency range: endothelial (Ae)—0.0095 – 0.021 Hz; neurogenic (An)—0.021 – 0.052 Hz; myogenic (Am)—0.052 – 0.145 Hz; respiratory-venular (Av)—0.145 – 0.6 Hz; pulse-cardial (Ac) —0.6 – 2 Hz (**Figure 1B**). The perfusion level (M) and the amplitudes of the blood flow modulation mechanisms were assessed in perfusion units (PU).

The functional activity of the tone-forming blood flow modulation mechanisms (endothelial, neurogenic, and myogenic) was evaluated as follows – the higher the vasomotion amplitude, the lower the tone, and vice versa, the lower the vasomotion amplitude, the higher the tone generated by this regulatory mechanism. If we take the zero amplitude as the longitudinal axis of a microvessel (LAV), and the maximum vasomotion amplitude as the vascular wall (**Figure 1B**), then the dependence of the microvessel lumen size on the vasomotion amplitude is clearly evident.

At the first stage of the study, the main objective was to assess the functional activity of resistive arterioles depending on the blood pressure level. At this stage, the subjects included 90 people (47 men and 43 women) divided into three groups. The control group (NT) consisted of 32 clinically healthy normotensive volunteers. The second group consisted of 32 patients with stage 1 essential AH (AH1). The third group included 26 patients with stage 2 AH (AH2). All patients with AH who were receiving antihypertensive therapy had their therapy discontinued 10–14 days prior to the study (washed out). In the rest of the patients, AH was newly diagnosed

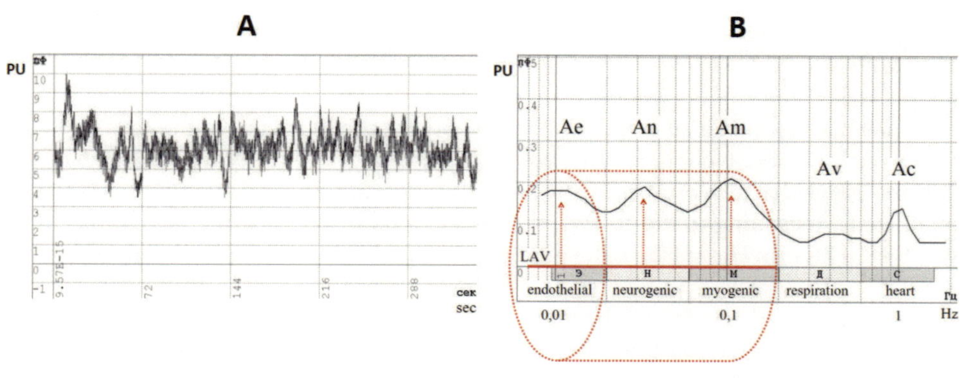

Figure 1.
Laser Doppler flowmetry (LDF). (A) It is a perfusion characteristic during 6 min. (B) It is amplitude-frequency wavelet analysis of blood flow oscillations. Dotted lines indicate a microvessel, arrows mark the activity of tone forming mechanisms in blood flow modulation. LAV, the longitudinal axis of the vessel.

	NT (n = 32)	AH1 (n = 32)	AH2 (n = 26)
Age (years)	48.9 ± 10.4	48.7 ± 11.2	49.8 ± 10.8
Sex (men/women)	13/19	17/15	17/9
primary/washed	– / –	20/12	9/17
SBP (mmHg)	117.5 ± 9.8	141.5 ± 12.8*	156.0 ± 15.4*,#
DBP (mmHg)	74.5 ± 8.6	86.8 ± 9.8*	95.2 ± 12.3*,#
HR (beats/min)	65.6 ± 6.9	68.2 ± 8.1	70.6 ± 9.3

*Differences are significant with respect to NT ($p < 0.000001$).
#Differences are significant with respect to AH1 ($p < 0.0001$).*

Table 1.
The main characteristics of the analyzed groups—first stage of the study.

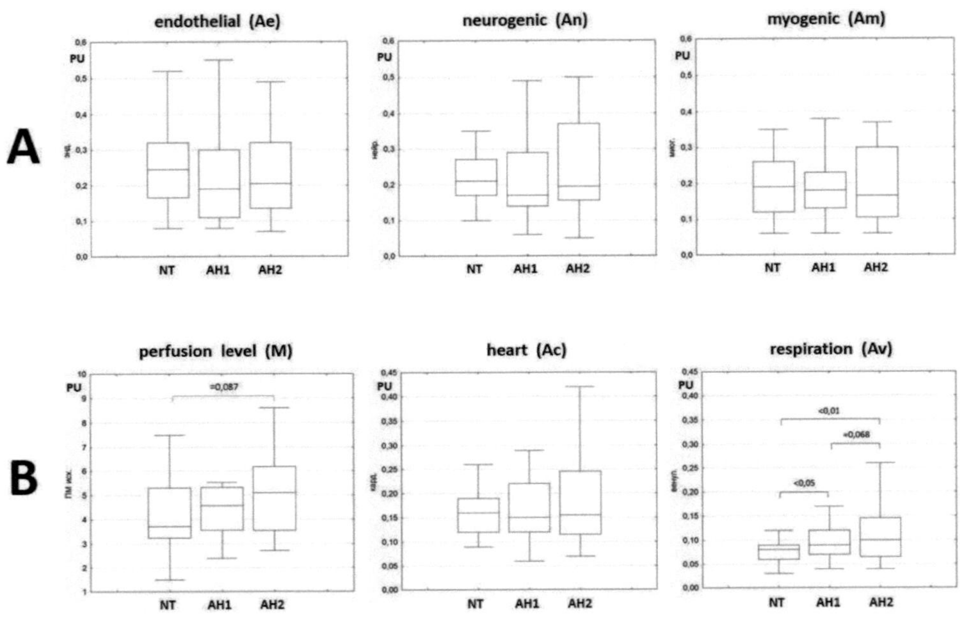

Figure 2.
LDF with amplitude-frequency wavelet analysis of blood flow in the first stage of the study. (A) The functional activity of tone-forming mechanisms (Ae, An, Am) of blood flow modulation. (B) The tissue perfusion (M) and the functional activity of the passive blood flow modulation mechanisms that reflect the condition of the blood inflow tracts to the capillary bed (Ac) and the outflow (Av) tracts. The rectangle indicates a range of 25–75 percentiles, and the median is indicated by a line.

and they had not received any drug therapy before the inclusion in the study (primary). The main characteristics of the analyzed groups and the hemodynamic parameters at the 10th minute of the adaptation period before LDF are presented in **Table 1**. According to the ABPM, the mean blood pressure (MBP) in the daytime was 89.8 ± 7.8 mm Hg in NT, 101.3 ± 5.2 mm Hg in AH 1, and 112.5 ± 6.9 mmHg in AH 2.

The functional activity of the main tone-forming microcirculation modulation mechanisms is shown in **Figure 2A**. Neither the expected increase in the basal myocyte tone (Am), nor a rise in sympathetic activity (An), nor signs of microvascular endothelial vasomotor dysfunction (Ae) were observed in any of the groups.

Figure 2B presents the analysis results of the tissue perfusion (M) and passive blood flow modulation mechanisms (Ac and Av) that determine the blood filling of the microvascular bed (MVB). The data obtained demonstrate that statistically significant differences are present only in blood volume in the venular section of the MVB (Av), which indicates a gradual increase in the venular blood volume as AH progresses. It is very important that out of all six parameters analyzed, only the amplitude of respiratory-associated blood flow oscillations (Av) had a weak but significant positive correlation with the MBP in the daytime ($r = 0.25$; $p = 0.035$) and night time ($r = 0.29$; $p = 0.013$). This correlation means that the more blood there is in the venular MVB, the higher the BP.

From the results obtained, it can be concluded that the functional state of the blood outflow tracts from the capillary bed plays a more significant role in the total peripheral vascular resistance (TPVR) than the functional state of resistive precapillary arterioles. The obtained data are in agreement with the opinion of Coulson et al. who distinguish two vascular resistance levels: the first is before the capillary plexus with an estimated contribution to TPVR of 67% according to the authors, and the second is after the capillary plexus with a contribution to TPVR of about 33% [34].

Functional disorders were dominant not in the blood inflow to the capillaries (resistive arterioles) but in the outflow system (venules), and that was a completely unexpected finding and does not fit into any of the existing AH development hypotheses (neurogenic, salt, membrane, etc.). It is also highly important that LDF demonstrated changes in tissue perfusion associated with respiratory movements of the chest in far from all AH patients.

And what is the nature of respiratory-associated tissue perfusion oscillations? The blood flow oscillations synchronous with breathing spread into microvessels from the capillary blood outflow side and are recorded in the venules. A mechanical passive transmission of respiratory intrathoracic pressure changes mediated by the venous system is discussed as their origin.

Normally, no respiratory-associated tissue perfusion oscillations are identified in LDF, regardless of the arterial blood volume flowing into the microcirculatory bloodstream. This is due to several factors. Firstly, the cross-sectional area and volume of the venular MVB significantly exceed the cross-sectional area and volume of the arteriolar MVB. Secondly, veins collapse. To maintain a round vein shape, a pressure of about 6–9 mmHg is required, and at lower values, the veins are ellipsoidal. With the same perimeter, the cross-sectional area of an ellipse is much smaller than that of a circle, so when the pressure increases from 0 to 6–9 mmHg, the capacity of the venous segment of the vascular bed increases substantially. Already at 10 mmHg, the increase in venous capacity is more than 60% of the maximum possible. Then the increase in the venous volume decelerates dramatically and is about 30% with a pressure rise from 10 to 80 mmHg [35].

How can intrathoracic pressure changes during respiration be conducted along the venous vasculature to the periphery, that is, to the postcapillary skin vessels? At low pressures, the collapsed venous walls will obviously dampen the retrograde respiratory wave propagation. Consequently, the propagation of the respiratory wave to the periphery can be observed only with fully expanded veins. Thus, the higher the blood volume in the venous bed, and hence the pressure in them, the better the respiratory waves are propagated to the periphery, and the higher is the Av amplitude. It is obvious that there is a certain critical Av value that reflects the degree of blood filling of the venous bed. During a long period of observation and measurements, the maximum Av value was found empirically to be 0.08 PU, which suggests that the veins are not yet fully expanded because the respiratory waves are not observed during the recording of tissue perfusion. At Av values of 0.09 PU, respiratory-associated oscillations of perfusion begin to appear in the LDF recording. These oscillations have a low amplitude and are not observed at every breath, which most likely depends on the volume and rate of respiratory movements.

	NT (n = 30)	VN (n = 30)	VP (n = 33)
Age (years)	44.9 ± 10.4	48.9 ± 10.3	47.8 ± 10.3
Sex (men/women)	15/15	17/13	17/16
primary/washed	– / –	12/18	11/22
SBP (mmHg)	118.0 ± 10.1	140.0 ± 14.1*	142.6 ± 14.7*
DBP (mmHg)	76.5 ± 9.0	88.3 ± 9.8*	91.1 ± 8.9*
HR (beats/min)	65.8 ± 8.5	68.1 ± 7.9	67.2 ± 9.1

**Differences are significant with respect to NT (p < 0.000005).*

Table 2.
The main characteristics of the analyzed groups—second stage of the study.

For the second stage of the study, a homogeneous group of 63 patients with AH was selected with mean daytime SBP of 140–159 mmHg according to ABPM and/or mean daily DBP of 90–99 mmHg. Just as at the first stage, all AH patients receiving antihypertensive therapy had their therapy discontinued 10–14 days prior to the study (washed out), or AH was diagnosed for the first time with no drug therapy before inclusion into the study (primary). The control group (NT) consisted of 30 clinically healthy normotensive volunteers. Based on the functional state of the venular microvessels (Av), the AH patients were divided into two groups. The first (VN) included 30 patients (48%) with no signs of an increased blood volume in the venular MVB (Av $\leq$ 0.08 PU). The second group (VP) consisted of 33 patients (52%) with signs of hypervolemia in the blood outflow tracts from the capillary bed (Av $\geq$ 0.09 PU) of varying severity. The groups characteristics and the hemodynamic parameters immediately before LDF (10th minute of the adaptation period) are presented in **Table 2**.

Based on the grouping parameter (Av) value, 8 subjects in the NT group (27%) had a moderate increase in blood volume in the venular MVB – Av = 0.09 PU (n = 3), Av = 0.1 PU (n = 4), Av = 0.12 PU (n = 1).

Figure 3 presents clinical examples of LDF in the groups analyzed. BP was measured 5 min before the start of LDF. **Figure 3B** shows that not every respiratory movement of a normotensive volunteer from the NT group is accompanied by a distinct change in tissue perfusion, and the main differences are observed in the amplitude of respiratory-associated blood flow oscillations. In a VP patient (**Figure 3D**), the differences are more pronounced not in amplitude but in the frequency of perfusion changes that coincide with the respiration frequency.

An analysis of the perfusion level and microcirculatory blood flow modulation mechanisms which determine the MVB blood filling demonstrated that in the VN group there was no statistically significant difference in tissue perfusion and pulse oscillation amplitude relative to NT, and the blood filling of venules is lower (**Figure 4A**). The situation is totally different in the VP group. A statistically significant rise in tissue perfusion relative to NT ($p < 0.0003$) can be explained by an increased contribution to the power of the signal reflected from red blood cells in the venular MVB. An unexpected finding was a significant increase in the amplitude of pulse oscillations relative to NT ($p < 0.004$), which indicates a higher arterial blood inflow to the exchange vessels. The obtained results demonstrate that the increase in the MVB perfusion is caused not only by a larger blood volume in the venular section (disrupted outflow) but also by an increased inflow of arterial blood.

An analysis of the functional activity of the tone-forming mechanisms demonstrated a significant ($p < 0.002$) decrease was also observed in the amplitude of

DOI: http://dx.doi.org/10.5772/intechopen.89852

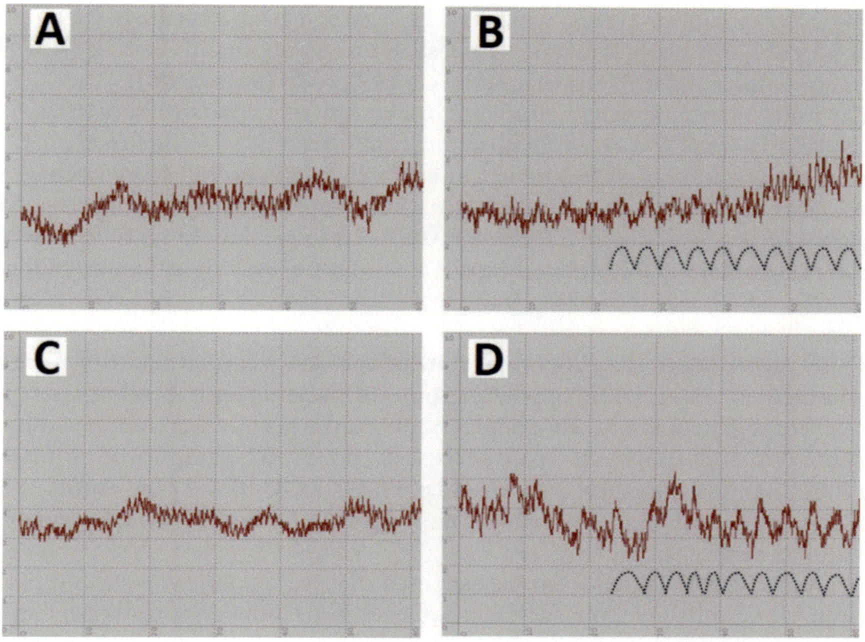

Figure 3.
The first minute of LDF. The functional state of the venular MVB. (A) NT group, BP is 115/70 mmHg: Respiratory-associated blood flow oscillations are not detected, perfusion (M) is 3.81 PU, respiration rate (RR) is 16/min, and Av is 0.05 PU. (B) NT group, BP is 130/80 mmHg: M – 3.82 PU, RR – 16/min, Av – 0.10 PU. (C) VN group, BP is 145/80 mmHg: M – 3.56 PU, RR – 16/min, Av – 0.07 PU. (D) VP group, BP is 150/90 mmHg: M – 3.89 PU, RR – 16/min, Av – 0.20 PU. The dotted lines reflect the changes in tissue perfusion synchronized with the respiratory chest movements.

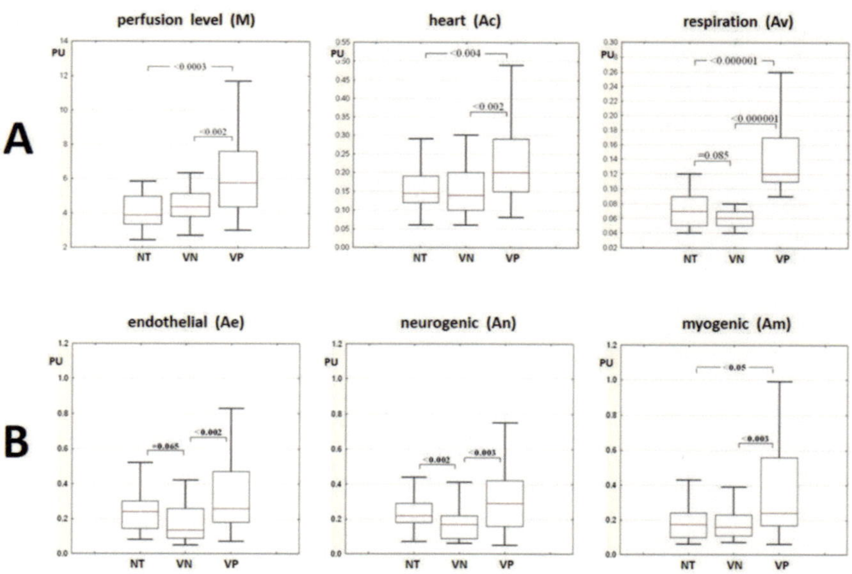

Figure 4.
LDF with amplitude-frequency wavelet analysis of blood flow in the second stage of the study. (A) The tissue perfusion (M) and the functional activity of the microcirculation modulation mechanisms which determine the blood filling of the microcirculatory bloodstream – The pulse (Ac) and respiratory-associated (Av) blood flow oscillation amplitude. (B) The functional activity of the tone-forming microcirculation modulation mechanisms (Ae, an, Am).

neurogenic vasomotions (An), which can be regarded as an increase in the sympathetic adrenergic tone. Meanwhile, the basal tone of smooth muscle cells (Am) was unchanged and comparable to that in the control group (**Figure 4B**).

The tone-forming mechanisms were in a completely opposite state in the VP group. Relative to NT, there were no significant differences in the functional activity of the endothelial and neurogenic arteriolar tone regulation mechanisms, although there was a trend towards a vasomotion amplitude increase by these regulatory mechanisms. However, the basal tone of the smooth muscle cells in precapillary arterioles and capillary sphincters was significantly reduced ($p < 0.05$), which was indicated by the myogenic vasomotion amplitude (Am) increase.

Thus, according to the functional state of the skin microcirculatory vessels, we obtained two completely opposite groups of patients with AH that initially seemed homogeneous. The differences between the groups were observed in all the six parameters analyzed: tissue perfusion ($p < 0.002$), amplitude of pulse ($p < 0.002$) and respiratory-associated ($p < 0.000001$) blood flow oscillations, amplitude of endothelial ($p < 0.002$), neurogenic ($p < 0.003$), and myogenic ($p < 0.003$) vasomotions.

From the data obtained it can be assumed that in AH patients without disruptions blood outflow from the capillary bed, hypertension is caused by an increase in the sympathetic adrenergic vascular tone regulation mechanism and by a vasomotor dysfunction of the microvascular endothelium. In patients with impaired blood outflow from the capillary bed associated with a decline in the myogenic tone of precapillary arterioles, an increase in the arterial blood inflow is observed.

The ABPM results were no less interesting. **Figure 5** shows significantly higher BP during both day and nighttime in the VP patients relative to VN, except for the nocturnal diastolic BP. A paradoxical situation arose – BP was higher in patients with a reduced resistive arteriolar tone than in patients with an increased tone.

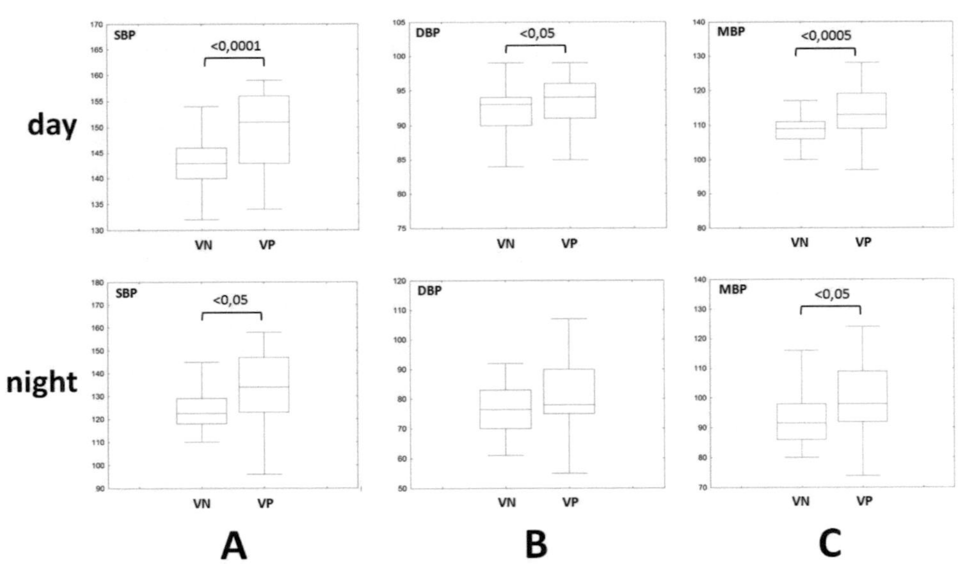

Figure 5.
The results of 24-h arterial blood pressure monitoring (ABPM). (A) Systolic BP (SBP). (B) Diastolic BP (DBP). (C) Mean BP (MBP).

3. Discussion

The pilot study of the functional state of the skin microvascular bed in essential arterial hypertension raised fundamental questions that require reflection and detailed consideration.

The first fundamental question is the validity of LDF in assessing the functional state of the resistive vascular bed in AH. On the one hand, there is no statistically significant correlation of BP with the functional activity of the tone-forming mechanisms in resistive arterioles. This can be explained by the absence of regulatory baroreflex mechanisms in the skin microvessels in contrast to microvessels in the striated muscle. On the other hand, the amplitude-frequency wavelet analysis of microcirculatory blood flow oscillations demonstrated that the functional state of the regulatory mechanisms in exchange microvessels may differ drastically in the initially homogeneous group of AH patients. This, in turn, not only has different hemodynamic and metabolic effects but quite evidently demands an individual approach to selecting antihypertensive therapy.

Let us consider the VN group in terms of the functional state of resistive arterioles. This group has a moderate decrease in the endothelial vasomotion amplitude with a trend towards significance ($p = 0.065$), indicating vasomotor dysfunction of the microvascular endothelium, which is consistent with the previously obtained results [36]. The group also exhibits a significant reduction in the neurogenic vasomotion amplitude ($p < 0.002$), which can be regarded as an increase in the sympathetic adrenergic tone. From the data obtained, it can be concluded that the functional state of resistive arterioles in AH patients with normal blood outflow from the capillary bed does not contradict Lang's neurogenic theory and the current ideas about the role of endothelial dysfunction in the AH pathogenesis. In such a functional state of the resistive arterioles, the prescription of drugs which lead to a tone decrease through various regulatory mechanisms can be considered quite justified.

Now let us consider the VP group, which raises many more questions. Based on the precapillary arteriole functional state, it can be concluded that there is a significant decrease in the basal tone of smooth muscle cells in arterioles and capillary sphincters. The insignificant (relative to NT) trend towards an amplitude increase (tone decline) of endothelial and neurogenic vasomotions can be explained in two ways: 1) changes in the regulatory systems themselves; 2) decreased smooth muscle cell sensitivity to the regulatory effects of the endothelium and the sympathetic adrenergic system. The hemodynamic consequences of the precapillary arteriolar tone are an increase in the arterial blood inflow (Ac), venular blood volume (Av), and as a result, tissue hyperperfusion, i.e. an increase in M.

Gryglewska et al. [37] also draw attention to a significant increase in the vasomotion amplitude in the neurogenic and myogenic activity range in patients with masked hypertension. The patients analyzed in that study were on average 10 years younger than our subjects. The data presented by the authors demonstrate that a neurogenic tone decline is observed against the background of an increased norepinephrine level. One possible cause for the increased plasma norepinephrine is the neurotransmitter leakage from the neuro-muscular synapses in microvessels in AH patients when the sympathetic nervous system activity is enhanced. This phenomenon is well-established for skeletal muscles [38, 39], but not for the skin microvessels which are not subject to baroreflex regulation [40]. The authors of the study suggest that the vasomotion amplitude increase in patients with masked hypertension is of a compensatory nature, aimed at meeting the metabolic needs of tissues when there is a reduced number of functioning capillaries, while the increase in blood pressure is due to the influence of various hormones, proinflammatory cytokines, and other humoral substances possibly secreted by the visceral adipose tissue [37].

Similar results for the functional state of the resistive skin microvessels in AH patients were also obtained also in another study [41]. In an age-matched group, *Rossi* et al. observed a significant increase in tissue perfusion compared to normotensive subjects, associated with significantly higher vasomotion amplitudes within the range of the endothelial, neurogenic, respiratory (venular), and cardial blood flow modulation mechanisms. The authors explain the enhanced vasomotor function of the microvascular endothelium as a response to an increase in the shear stress at high BP values, and the higher vasomotion amplitude in the range of the neurogenic regulation mechanism as sympathetic nervous system activation. The authors consider the observed increase in the pulse oscillation amplitude a manifestation of a high BP level, and the rise in the respiratory-associated blood flow oscillation amplitude as a compensatory reaction aimed at extending the time for oxygen extraction by tissues under microvascular rarefication.

Hemodynamic changes at the MVB level entail significant metabolic changes. Of the three main metabolic mechanisms (diffusion, filtration-reabsorption, and vesicular), filtration-reabsorption depends directly on the hemodynamic parameters. This metabolic mechanism is described by the Starling equation and provides bidirectional transendothelial transport of water-soluble and low-molecular-weight substances due to the difference in the hydrostatic and colloid osmotic blood pressures. Enhanced arterial blood inflow (Ac) with impaired venular outflow (Av) in VP patients can lead to a substantial rise in hydrostatic pressure in the capillaries, which is observed in far from all AH patients during direct pressure measurement in the nail bed capillaries [42–44]. An increase in the capillary hydrostatic pressure leads to a shift in the filtration-reabsorption metabolic mechanism towards the predominance of filtration, with developing hyperhydration of the interstitial space and latent fluid retention in the tissues [45, 46]. The hyperhydration of the interstitial space increases the blood↔cell distance for nutrients and tissue metabolism products, leading to a disruption in the metabolic process rate.

Another very important issue is the choice of antihypertensive therapy in this group of patients. The choice of vasodilating drugs is highly dubious. An additional decline in the initially low resistive arteriolar tone can have a range of negative consequences, including orthostatic instability, syncope, peripheral edema, etc. The study by Makolkin et al. [47] is highly illustrative in this respect. The authors define several hemodynamic microcirculation types (normocirculatory, spastic, hyperemic, and congestive-stasical) and note the greatest effect of calcium antagonists in the spastic type, while these cause puffiness in the lower legs and feet, as well as hot flashes in the congestive-stasical type.

It can be speculated that patients in this group are poorly compliant with therapy due to low tolerability, or it is difficult to achieve the target BP. As Engholm et al. demonstrated in their study, in patients with a low systemic vascular resistance index (SVRI), additional administration of antihypertensive drugs from other groups (two-component therapy) significantly lowers BP without changing SVRI. There is also an insignificant trend towards an increase in the left atrial size with a decrease in the left ventricular stroke volume and cardiac index relative to patients with high SVRI [48]. Based on the functional state of the arteriolar and venular microvessels in the VP group, it can be assumed that in this case the prescription of drugs with venotonic and/or diuretic effects will be most justified.

The next important question is the reasons for such a substantial difference in the functional state of the skin microvessels, when all the parameters analyzed by LDF differ significantly. It can be assumed that this is due either to different etiological factors or to the pathological process duration. The early AH development stages are not marked by elevated levels of endothelin, a marker of vasomotor endothelial dysfunction [49]. We were not able to assess the pathological process duration in the analyzed groups, which is one among many weaknesses of this pilot study, but it can be assumed, based on the significant reduction in the microvascular endothelial vasomotor function, that the pathological process had been developing longer in the VN patients.

The role of the temporal factor in the detected functional differences is supported by studies of vasomotor function of microvessels in different AH models in laboratory animals. Boegehold et al. demonstrated that a high-salt diet for 6–7 weeks in healthy four-week old rats resulted in a significantly higher amplitude of vasomotions in the mesenteric arterioles with a diameter of 30 μm compared to the control group that did not received the salt load [50]. In another study in 10–12 week-old spontaneously hypertensive rats (SHR), Noble et al. not only observed a significant rise in the vasomotion amplitude in the spinal skeletal muscle arterioles with a diameter of <30 μm relative to those in normotensive animals, but also a significant increase in the diameter of precapillary arterioles (6–15 μm) and postcapillary venules (15–40 μm) [51]. In another study of the cerebral microvasculature in SHR, a significant increase in the vasomotion amplitude was demonstrated in micro- and larger (diameter 30–70 μm) vessels, where the smooth muscle layer is more pronounced and the neurogenic (sympathetic adrenergic) vascular tone regulation mechanism predominates [52]. The obtained experimental results in laboratory animals are quite consistent with the functional state of microvessels in the VP group, which may indirectly indicate the initial manifestations of the pathology in this group of patients.

This hypothesis is also supported by the high correlation of Av with MBP obtained in the combined NT + VP group ($n = 63$): day time – $r = 0.66$ ($p < 0.0001$); night time – $r = 0.76$ ($p < 0.0001$). This relationship indicates that the progression of disorders in the venular-venous vascular bed is accompanied by an increase of BP. It can be hypothesized that first the renin-angiotensin system is activated to maintain tissue homeostasis and compensate the filtration-reabsorption imbalance, leading to an increase in the resistive arteriolar tone, which in turn leads to a decrease in hydrostatic blood pressure in the capillaries. This regulatory mechanism restores the filtration-reabsorption metabolic balance but triggers a vicious circle which ultimately leads to structural microvascular changes. Based on this hypothesis, it can be suggested that the VP patients are at the early AH stage (functional impairment stage), and the VN patients have a longer course of the pathological process with the development of structural changes. It is extremely difficult to determine the AH course duration since many patients do not feel elevated BP. In this study, in 23 patients (37%) from both groups AH was diagnosed incidentally without any previous suspicion on their part about the existence of a pathology.

Another important issue concerns the causes of disrupted blood outflow from the capillary bed, which may be due to several factors or their combination. One may be a disturbance in the functional state of the major veins which is observed both in laboratory animals [51] and in some AH patients and is expressed as a decrease in the venous wall elasticity with an elevation of their tone and the blood pressure in them [53–56]. Another cause may be an increase in the right atrial pressure, which is observed in some AH patients [57, 58].

Another highly important issue concerns the BP level. What causes a higher BP level in VP patients relative to the VN group? The venular microvessels are generally believed to contribute not more than 8–9% of the PVR [59], although, as noted above, the contribution of the postcapillary section may be much higher [34]. But here another factor should be considered that significantly affects BP, i.e. the cardiac output. The cardiac output in AH patients may be elevated during venoconstriction [60]. Based on the significantly lower peripheral vascular resistance, the VP patients can be hypothesized to have a hyperkinetic circulation type, and their higher BP values are due to a combination of two factors – an increase in the cardiac output, as noted in other studies [48], and a disruption of the blood outflow from the capillary bed with an increase in PVR at the postcapillary (venular) level.

Regardless of the reasons for the hindered blood outflow from the capillary bed, the mechanism that increases the amplitude of myogenic vasomotions in the precapillary arterioles and capillary sphincters is baffling. Normally, in a closed cardiovascular system, an increase in the venous pressure results in a consecutive rise in the capillary and precapillary arteriolar pressure. The smooth muscle cells of the resistive microvessels respond to increased pressure and distension by contracting by the Ostroumov-Bayliss mechanism, which should lead to a rise in their tone with a decrease in the amplitude of myogenic vasomotions (Am). But in the VP group, the exact opposite effect is observed. How can we explain it?

It can be hypothesized that the precapillary arteriolar tone decline is of a compensatory nature, aimed at overcoming the raised resistance at the postcapillary level. The increase in the arterial blood inflow is aimed at overcoming the impaired venous outflow. But from the metabolic expediency point of view, this compensation mechanism is extremely unfavorable. This can be assumed to be a consequence of the functional features of the smooth muscle cells in VP patients. Falcone et al. established *in vivo* that an increase in venous pressure in the cremaster muscle of SHR caused a more pronounced constriction reaction of the 15–100 μm arterioles than in normotensive animals [61]. Another research group demonstrated on mesenteric arterioles (diameter 100–150 μm) *in vitro* that in SHR the smooth muscle cells (lacking sympathetic innervation) develop a significantly higher vasomotion amplitude in response to norepinephrine compared to normotensive animals [62].

The results of this pilot study suggest that the contribution of the venular MVB to PVR is more substantial, and the generally accepted vascular resistance scale requires adjustment. In previous studies on healthy normotensive volunteers it was demonstrated that the blood volume in the skin venular microvessels has a positive correlation with BP, and a negative correlation not only with the magnitude of nocturnal BP decline, but also with the dilation response of MVB during the heat test, the nociception system activation, and post-occlusive reactive hyperemia [63, 64]. MVB is anatomically located between the arterial and venous systems, and, as capillaries do not stretch, the disruptions in the outflow system naturally affect the inflow system to the exchange vessels with all the logical metabolic consequences. This state can be formulated simply as "no outflow – no inflow".

The venular MVB itself deserves special attention. What is the physiological underpinning of the significantly larger vascular volume in the venular MVB than on the arteriolar side? This is what Nature intended, and therefore there is a physiological reason. Unlike arterioles, which only regulate blood inflow to the capillaries, many important functions are performed at the venular level. One is that mast cells are located in the immediate vicinity of venules with a diameter of 40–80 μm. This protective system reacts to various chemical and biological noxious agents entering the systemic circulation and it is situated in the venular section of the MVB. Postcapillary venules provide the exchange of macromolecules and protein-bound substances in the bloodstream by vesicular transport. Recognition, immobilization, "unpacking" of transport molecules, and vesicle formation require some time and are facilitated by the hemodynamic conditions in the venular section of MVB, since it has the slowest blood flow in the entire cardiovascular system. The venular section occupies a strategic position and is the first to receive metabolic information.

In the conducted studies it was discovered that any changes in venous MVB affect the functional state of the adjoining afferent arteriole and the effects are observed up to 12 mm from the stimulus applied. The following was revealed: constriction reactions of arterioles to intravenular administration of norepinephrine [65], dilation reactions to acetylcholine [66], adenosine [67], ATP [68], dilator prostanoids [69], and even to the metabolic stimulator 2,4-dinitrophenol [70]. In response to an increase in shear stress, venular endothelium produces NO leading to a corresponding dilation of the adjoining arteriole [71]. There is an opinion that the arteriolar NO level is determined by its venular concentration, whose basal level is maintained by the secretory activity of the venular endothelium [72]. The endothelium of the venous bed is also sensitive to the blood gas composition. Tkachenko demonstrated that with blood O_2 saturation < 68% sympathetic stimulation leads to a contraction of the muscle veins (>100–150 μm), when O_2 is in the range of 68–85%, the reaction can be both constriction and dilatation, but at saturation values of >85%, sympathetic stimulation causes only venous dilation [73].

4. Conclusion

LDF is a relatively new technique for studying microcirculation in humans which is going through the stages of adoption, data accumulation, and interpretation of the obtained results. A review of the scientific literature over the past 20 years has identified only 143 studies of microcirculation in humans by LDF with wavelet analysis of blood flow oscillation. The total number of healthy subjects and patients with various organ and system diseases involved in the studies does not exceed 2600 [74].

The main aim of this pilot study was to evaluate the informativeness of LDF in assessing the functional state of the skin microvascular bed in patients with essential arterial hypertension. The study has many weaknesses, due to the retrospective data analysis and the inability to collect information on the central hemodynamic parameters and humoral status of the patients. Nevertheless, it can be concluded that the aim has been achieved. The obtained results are unexpected; however, they allow the microcirculatory blood flow problems in patients with AH to be viewed in an entirely new light.

Several quite precise conclusions can be made based on the study:

1. LDF is a very promising method to study the functional state of microcirculation in humans and can be useful for the selection of personalized antihypertensive therapy.

2. The functional activity of the tone-forming mechanisms (endothelial, neurogenic, myogenic) of the resistive arterioles in the skin is not correlated with BP according to ABPM data.

3. AH patients can differ significantly in their functional status of resistive microvessels. One half of the patients have an increased arteriolar tone without hemodynamic disruptions in the venular MVB. The other half of the patients is a complete opposite, that is, they exhibit a decrease in arteriolar tone associated with disrupted blood outflow from the capillaries. These functional differences must be considered when selecting antihypertensive therapy. This may assist in increasing the effectiveness of therapy and patient compliance with the treatment.

4. The role of the venular-venous vascular bed in AH is underestimated, and further studies in this area are required; this will contribute to solving many problems that cardiologists face today.

Acknowledgements

The author is grateful to the employees of the Federal State Institution, National Medical Research Center for Cardiology of the Ministry of Healthcare of the Russian Federation Sergey A. Boytsov, Anatoly N. Rogoza, Shurat B. Gorieva, Tatyana S. Pavlova, Marina V. Sergeeva for their help in recruiting clinical material and the Facecontrol, Inc. for financial support.

Conflict of interest

The author declare no conflict of interests.

Author details

Andrey A. Fedorovich[1,2]

1 Federal State Institution, National Medical Research Center for Preventive Medicine of the Ministry of Healthcare of the Russian Federation, Moscow, Russian Federation

2 State Scientific Center of Russia, Institute of Biomedical Problems of the Russian Academy of Sciences, Moscow, Russian Federation

*Address all correspondence to: faa-micro@yandex.ru

IntechOpen

© 2019 The Author(s). Licensee IntechOpen. This chapter is distributed under the terms of the Creative Commons Attribution License (http://creativecommons.org/licenses/by/3.0), which permits unrestricted use, distribution, and reproduction in any medium, provided the original work is properly cited.

References

[1] Levy BI, Ambrosio G, Pries AR, Struijker-Boudier HA. Microcirculation in hypertension. A new target for treatment? Circulation. 2001;**104**(6): 735-740. DOI: 10.1161/hc3101.091158

[2] Feihl F, Liauder L, Waeber B, Levy BI. Hypertension a disease of the microcirculation? Hypertension. 2006;**48**(6):1012-1017. DOI: 10.1161/01.HYP.0000249510.20326.72

[3] Eftekhari A, Mathiassen ON, Buus NH, Gotzsche O, Mulvany MJ, Christensen KL. Disproportionally impaired microvascular structure in essential hypertension. Journal of Hypertension. 2012;**30**(4):794-801. DOI: 10.1097/HJH.0b013e3283447a1c

[4] Blacher J, Evans A, Arveiler D, Amouyel P, Ferrieres J, Bingham A, et al. Residual cardiovascular risk in treated hypertension and hyperlipidaemia: The PRIME study. Journal of Human Hypertension. 2010;**24**(1):19-26. DOI: 10.1038/jhh.2009.34

[5] Rossi M, Taddei S, Fabbri A, Tintori G, Credidio L, Virdis A, et al. Cutaneous vasodilation to acetylcholine in patients with essential hypertension. Journal of Cardiovascular Pharmacology. 1997;**29**(3):406-411. DOI: 10.1097/00005344-199703000-00015

[6] Shamin-Uzzaman QA, Pfenninger D, Kehrer C, Chakrabarti A, Kacirotti N, Rubenfire M, et al. Altered cutaneous microvascular responses to reactive hyperemia in coronary artery disease: A comparative study with conduit vessel responses. Clinical Science. 2002;**103**(3):267-273. DOI: 10.1042/cs1030267

[7] Stewart J, Kohen A, Brouder D, Rahim F, Adler S, Garrick R, et al. Noninvasive interrogation of microvasculature for signs of endothelial dysfunction in patients with chronic renal failure. American Journal of Physiology. Heart and Circulatory Physiology. 2004;**287**(6):H2687-H2696. DOI: 10.1152/ajpheart.00287.2004

[8] Holovatz LA, Thompson-Torgerson CS, Kenney WL. The human cutaneous circulation as model of generalized microvascular function. Journal of Applied Physiology. 2008;**105**(1):370-372. DOI: 10.1152/japplphysiol.00858.2007

[9] Dubiel M, Krolczyk J, Gasowski J, Grodzicki T. Skin microcirculation and echocardiographic and biochemical indices of left ventricular dysfunction in non-diabetic patients with heart failure. Journal of Cardiology. 2011;**18**(3): 270-276. PMID: 21660916

[10] Hansell J, Henareh L, Agewall S, Norman M. Non-invasive assessment of endothelial function – Relation between vasodilatory responses in skin microcirculation and brachial artery. Clinical Physiology and Functional Imaging. 2004;**24**(6):317-322. DOI: 10.1111/j.1475-097X.2004.00575.x

[11] Coulon P, Constans J, Gosse P. Impairment of skin blood flow during post-occlusive reactive hyperhemy assessed by laser Doppler flowmetry correlates with renal resistive index. Journal of Human Hypertension. 2012;**26**(1):56-63. DOI: 10.1038/jhh.2010.117

[12] Sieg-Dobrescu D, Burnier M, Hayoz D, Brunner H-R, Waeber B. The return of increased blood pressure after discontinuation of antihypertensive treatment is associated with an impaired post-ischemic skin blood flow response. Journal of Hypertension. 2001;**19**(8):1387-1382. PMID: 11518846

[13] Vuilleumitr P, Decosterd D, Maillard M, Burnjer M, Hayoz D.

Postischemic forearm skin reactive hyperemia is related to cardiovascular risk factors in a healthy female population. Journal of Hypertension. 2002;**20**(9):1753-1757. DOI: 10.1097/00004872-200209000-00018

[14] Lekakis J, Abraham P, Balbarini A, Blann A, Boulanger CM, Cockroft J, et al. Methods for evaluating endothelial function: A position statement from the European Society of Cardiology Working Group on peripheral circulation. European Journal of Cardiovascular Prevention and Rehabilitation. 2011;**18**(6):775-789. DOI: 10.1177/1741826711398179

[15] Frohlich E.D., Ventura H.O. Pathophysiology: disease mechanisms. In: The Hand Book "Hypertension": An Atlas of Investigation and Management. Clinical Publishing Oxford, 2009:1-15

[16] Zweifach BW. Quantitative studies of microcirculatory structure and function. I. Analysis of pressure distribution in the terminal vascular bed in cat mesentery. Circulation Research. 1974;**34**(6):843-857. DOI: 10.1161/01.RES.34.6.841

[17] Zweifach BW. Quantitative studies of microcirculatory structure and function. II. Direct measurement of capillary pressure in splanchnic mesenteric vessels. Circulation Research. 1974;**34**(6):858-866. DOI: 10/1161/01.RES.34.6.858

[18] Caro CG, Pedly TJ, Schroter RC, Seed WA. The Mechanics of the Circulation. Hand Book. Cambridge: University Press; 2012. p. 550. DOI: 10.1017/CBO9781139013406

[19] Braverman IM. The cutaneous microcirculation: Ultrastructure and microanatomical organization. Microcirculation. 1997;**4**(3):329-340. DOI: 10.3109/10739689709146797

[20] Chambers R, Zweifach BW. Functional activity of blood capillary bed, with special reference to visceral tissue. Annals of the New York Academy of Sciences. 1944;**46**(8):683-694. DOI: 10.1111/j.1749-6632.1946.tb31697.x

[21] Meyer MF, Rose CJ, Hülsmann J-O, Schatz H, Pfohl M. Impaired 0.1-Hz vasomotion assessed by laser Doppler anemometry as an early index of peripheral sympathetic neuropathy in diabetes. Microvascular Research. 2003;**65**(2):88-95. DOI: 10.1016/S0026-2862(02)00015-8

[22] Krupatkin AI, Sidorov VV, Fedorovich AA, Efimochkin SA, Zeinalov VT. The oscillatory circuit for the control of functional capillaries number. Regional Blood Circulation and Microcirculation. 2006;**21**(3):54-58 (in Russian)

[23] Funk W, Intaglietta M. Spontaneous arteriolar vasomotion. Progress in Applied Microcirculation. 1983;**3**:66-82. DOI: 10.1159/000409287

[24] Kastrup J, Bulow J, Lassen NA. Vasomotion in human skin before and after local heating recorder with laser Doppler flowmetry. International Journal of Microcirculation. 1989;**8**(2):205-215. PMID: 2659545

[25] Bertuglia S, Colantuoni A, Coppini G, Intaglietta M. Hypoxia- or hyperoxia-induced changes in arteriolar vasomotion in skeletal muscle microcirculation. American Journal of Physiology. 1991;**260**(2Pt2):H362-H372. DOI: 10.1152/ajpheart.1991.260.2.H362

[26] Bollinger A, Yanar A, Hoffmann U, Franzeck UK. Is high-frequency flux motion due to respiration or to vasomotion activity? In: Allegra C, Intaglietta M, Messmer K, editors. Progress in Applied Microcirculation. Vol. 20. Basel, Karger; 1993. pp. 52-58. DOI: 10.1159/000422452

[27] Colantuoni A, Bertuglia S, Intaglietta M. Microvascular vasomotion: Origin of laser Doppler fluxmotion. International Journal of Microcirculation, Clinical and Experimental. 1994;**14**(3):151-158. DOI: 10.1159/000178823

[28] Parthimos D, Edwards DH, Griffith TM. Comparison of chaotic and sinusoidal vasomotion in the regulation of microvascular flow. Cardiovascular Research. 1996;**31**(3):388-399. DOI: 10.1016/S0008-6363(95)00123-9

[29] Schmid-Schönbein H, Ziege S, Rütten W, Heidtmann H. Active and passive modulation of cutaneous red cell flux as measured by laser Doppler anemometry. VASA. 1992;**34**(Suppl): 38-47. PMID: 1388307

[30] Schmid-Schönbein H, Ziege S, Grebe R, Blazek V, Spielmann R, Linzenich F. Synergetic interpretation of patterned vasomotor activity in microvascular perfusion: Discrete effects of myogenic and neurogenic vasoconstriction as well as arterial and venous pressure fluctuations. International Journal of Microcirculation, Clinical and Experimental. 1997;**17**(6):349-359. DOI: 10.1159/000179251

[31] Muck-Weymann ME, Albrecht HP, Hiller D, Hornstein OP, Bauer RD. Respiration-dependence of cutaneous laser Doppler flow motion. VASA. 1994;**23**(4):299-304. PMID: 7817609

[32] Stefanovska A, Bracic M, Kvernmo HD. Wavelet analysis of oscillations in peripheral blood circulation measured by Doppler technique. IEEE Transactions on Biomedical Engineering. 1999;**46**(10):1230-1239. PMID: 10513128

[33] Bernjak A, Clarkson PBM, McClintock PVE, Stefanovska A. Low-frequency blood flow oscillations in congestive heart failure and after β_1-blocade treatment. Microvascular Research. 2008;**76**(3):224-232. DOI: 10.1016/j.mvr.2008.07.006

[34] Coulson RL, Grayson J, Irvine M. Observations on coronary collateral communications and the control of flow in the coronary circulation of the dog. The Journal of Physiology. 1970;**208**(3):563-581. DOI: 10.1113/jphysiol.1970.sp009137

[35] Oberg B. The relationship between active constriction and passive recoil of the veins at various distending pressures. Acta Physiologica Scandinavica. 1967;**71**(2):233-247. DOI: 10.1111/j.1748-1716.1967.tb03729.x

[36] Farkas K, Kolossvary E, Jarai Z, Nemcsik J, Farsang C. Non-invasive assessment of microvascular endothelial function by laser Doppler flowmetry in patients with essential hypertension. Aterosclerosis. 2004;**173**(1):97-102. DOI: 10.1016/j.atherosclerosis.2003.11.015

[37] Gryglewska B, Necki M, Cwynar M, Baron T, Grodzicki T. Neurogenic and myogenic resting skin blood flow motion in subjects with masked hypertension. Journal of Physiology and Pharmacology. 2010;**61**(5):551-558. PMID: 21081798

[38] Grassi G, Seravalle G, Trevano FQ, Dell'oro R, Bolla G, Cuspidi C, et al. Neurogenic abnormalities in masked hypertension. Hypertension. 2007;**50**(3):537-542. DOI: 10.1161/HYPERTENSIONAHA.107.092528

[39] Wallin BG, Charkoudian N. Sympathetic neural control of integrated cardiovascular function: Insights from measurement of human sympathetic nerve activity. Muscle & Nerve. 2007;**36**(5):595-614. DOI: 10.1002/mus.20831

[40] Grassi G, Colombo M, Seravalle G, Spaziani D, Mancia G. Dissociation

between muscle and skin sympathetic nerve activity in essential hypertension, obesity and congestive heart failure. Hypertension. 1998;**31**(1):64-67. DOI: 10.1161/01.HYP.31.1.64

[41] Rossi M, Bradbury A, Magagna A, Pesce M, Taddei S, Stefanovska A. Investigation of skin vasoreactivity and blood flow oscillations in hypertensive patients: Effect of short-term antihypertensive treatment. Journal of Hypertension. 2011;**29**(8):1569-1576. DOI: 10.1097/HJH.0b013e328348b653

[42] Eichna LW, Bordley J. Capillary blood pressure in man. Direct measurements in the digits of normal and hypertensive subjects during vasoconstriction and vasodilatation variously induced. Journal of Clinical Investigation. 1942;**21**(6):711-729. DOI: 10.1172/JCHI101347

[43] Tooke JE, Williams SA. Capillary blood pressure. Advances in Experimental Medicine and Biology. 1987;**220**:209-214. PMID: 3673766

[44] Williams SA, Boolell M, MacGregor GA, Smaje LH, Wasserman SM, Tooke JE. Capillary hypertension and abnormal pressure dynamics in patients with essential hypertension. Clinical Science (London, England). 1990;**79**(1):5-8. DOI: 10.1042/cs0790005

[45] Kanishcheva E, Fedorovich A, Loukianov M, Boytsov S. Capillary nail bed parameters in hypertensives and normotensives in age group of 60-80 years. Journal of Hypertension. 2010;**28**(Suppl A):E182

[46] Gurfinkel YI, Makeeva OV, Ostrozhinsky VA. Features of microcirculation, endothelial function and pulse wave velocity in patients with early stages of hypertension. Functional Diagnostics. 2010;**2**:18-25 (in Russian)

[47] Makolkin VI. Microcirculation in hypertension. In: The hand book "Microcirculation in cardiology". Moscow: Vizart. pp. 88-112 (in Russian)

[48] Engholm M, Mulvany MU, Eftekhari A, Mathiassen ON, Buus NH, Christensen KL. Positive effects of aggressive vasodilator treatment of well-treated essential hypertensive patients. Journal of Human Hypertension. 2016;**30**(11):690-696. DOI: 10.1038/jhh.2016.13

[49] Taddei S, Virdis A, Ghiadoni L, Sudano I, Magagna A, Salvetti A. Role of endothelin in the control of peripheral vascular tone in human hypertension. Heart Failure Reviews. 2001;**6**:277-285. PMID: 11447302

[50] Boegehold MA. Enhanced arteriolar vasomotion in rats with chronic salt-induced hypertension. Microvascular Research. 1993;**45**:83-94. DOI: 10.1006/mvre.1993.1008

[51] Noble JLML, Smith TL, Hutchins PM, Struyker-Bodier AJ. Microvascular alterations in adult conscious spontaneously hypertensive rats. Hypertension. 1990;**15**:415-419. PMID: 2318521

[52] Lefer DJ, Lynch CD, Lapinski KC, Hutchins PM. Enhanced vasomotion of cerebral arterioles in spontaneously hypertensive rats. Microvascular Research. 1990;**39**:129-139. PMID: 2352485

[53] Walsh IA, Hyman CH, Maronde RF. Venous distensibility in essential hypertension. Cardiovascular Research. 1969;**3**:338-349. DOI: 10.1093/cvr/3.3.338

[54] Takeshita A, Mark A. Decreased venous distensibility in borderline hypertension. Hypertension. 1979;**1**:202-206. PMID: 551074

[55] Safar ME, London GM. Venous system in essential hypertension. Hypertension. 1985;**69**:497-504. DOI: 10.1042/cs0690497

[56] Delaney EP, Young CN, DiSabatino A, Stillabower ME, Farquhar WB. Limb venous tone and responsiveness in hypertensive humans. Journal of Applied Physiology. 2008;**105**:894-901. DOI: 10.1152/japplphysiol.90574.2008

[57] London GM, Safar ME, Simon AC, Alexandre JM, Levenson JA, Weiss YA. Total effective compliance, cardiac output and fluid volumes in essential hypertension. Circulation. 1978;**57**:995-1000. PMID: 639223

[58] Safar M, Plante G, London G. Vascular compliance and blood volume in essential hypertension. In: Lagard JH, Brenner BM, editors. Hypertension. New York: Raven Press Ltd; 1995. pp. 377-388

[59] Davis MJ, Ferrer PN, Gore RW. Vascular anatomy and hydrostatic pressure profile in the hamster cheek pouch. American Journal of Physiology. 1986;**50**(2 Pt2):H291-H303. DOI: 10.1152/ajpheart.1986.250.2.H291

[60] Schmieder RE, Messerli FH, Nunez BD, Garavaglia GE, Frohlich ED. Hemodynamic, humoral and volume findings in systemic hypertension with isolated ventricular septal hypertrophy. The American Journal of Cardiology. 1988;**15**:1053-1057. PMID: 2973218

[61] Falcone JC, Granger HJ, Meininger GA. Enhanced myogenic activation in skeletal muscle arterioles from spontaneously hypertensive rats. American Journal of Physiology. 1993;**265**(6Pt2):H1847-H1855. DOI: 10.1152/ajpheart.1993.265.6.H1847

[62] Chen X, Yang D, Ma S, He H, Luo Z, Feng X, et al. Increased rhythmicity in hypertensive arterial smooth muscle is linked to transient receptor potential canonical channels. Journal of Cellular and Molecular Medicine. 2010;**14**(10):2483-2494. DOI: 10.1111/j.1582-4934.2009.00890.x

[63] Fedorovich AA. Relationship of the functional state of arteriolar and venular parts of the vascular bed of the skin to the level of blood pressure. Regional Blood Circulation and Microcirculation. 2009;**32**(4):47-53 (in Russian)

[64] Fedorovich AA, Rogoza AN, Boitsov SA. Correlation of the arteriolar and venule segment functions of the vascular bed with dilatory reserve and central hemodynamics parameters. Functional Diagnostics. 2009;**1**:14-22 (in Russian)

[65] Tigno XT, Ley K, Pries AR, Haehtgens P. Venulo-arteriolar communication and propagated response. A possible mechanism for local control of blood flow. Pflügers Archiv (European Journal of Physiology). 1989;**414**:450-456. PMID: 2798041

[66] Falcone JC, Bohlen HG. EDRF from rat intestine and skeletal muscle venules causes dilation of arterioles. American Journal of Physiology. Heart and Circulatory Physiology. 1990;**258**:H1515-H1523. DOI: 10.1152/ajpheart.1990.258.5.H1515

[67] Hester RL. Venular-arteriolar diffusion of adenosine in hamster cremaster microcirculation. American Journal of Physiology. Heart and Circulatory Physiology. 1990;**258**:H1918-H1924. DOI: 10.1152/ajpheart.1990.258.6.H1918

[68] Hammer LW, Ligon AL, Hester RL. ATP-mediated release of arachidonic acid metabolites from venular endothelium causes arteriolar dilation. The American Journal of Physiology (Heart and Circulatory

Physiology). 2001;**280**:H2616-H2622. DOI: 10.1152/ajpheart.2001.280.6.H2616

[69] McKay MK, Gardner AL, Boyd D, Hester R. Influence of venular prostaglandin release on arteriolar diameter during functional hyperemia. Hypertension. 1998;**31**(2):213-217. DOI: 10.1161/01.HYP.31.1.213

[70] Saito Y, Eraslan A, Hester RL. Importance of venular flow in control of arteriolar diameter in hamster cremaster muscle. American Journal of Physiology. Heart and Circulatory Physiology. 1993;**265**:H1294-H1300. DOI: 10.1152/ajpheart.1993.265.4.H1294

[71] Boegehold MA. Shear-dependent release of venular nitric oxide: effect on arteriolar tone in rat striated muscle. The American Journal of Physiology (Heart and Circulatory Physiology). 1996;**271**:H387-H1395. DOI: 10.1152/ajpheart.1996.271.2.H387

[72] Nellore K, Harris NR. Nitric oxide measurements in rat mesentery reveal disrupted venulo-arteriolar communication in diabetes. Microcirculation. 2004;**11**:415-423. DOI: 10.1080/10739680490457809

[73] Tkachenko BI. Venous circulation. Medicine:Leningrad. 1974:224 (in Russian)

[74] Martini R, Bagno A. The wavelet analysis for the assessment of microvascular function with the laser Doppler fluxmetry over the last 20 years. Looking for hidden informations. Clinical Hemorheology and Microcirculation. 2018;**70**(2): 213-229. DOI: 10.3233/CH-189903